QUANTUM PROGRESSION™
The Art & Science of Career Advancement in The Age of A.I.

Valerie Capers Workman, Esq.

Former Vice President, People at Tesla and
Chief Legal Officer for Handshake

Dedication

To everyone who fears the future, please know the future is always now. You are already here. If you've made it this far, you can go further.

To my parents, James and Lucille, who made a veritable feast from leftovers.

To my husband, Antonio. We did it! I love us.

To our sons, Marcus, Desmond, and Rawle, the students who have become our teachers.

To my sisters, Gwen and Lily. Onward and upward!

THANK YOU, LORD!

TABLE OF CONTENTS

QUANTUM PROGRESSION™
The Art & Science of Career
Advancement in The Age of A.I.

Foreword

Artificial Intelligence is Inevitable: Resistance is Futile

I had been staring at the blank page for a good ten minutes. The task seemed simple enough—create a detailed job description for a new role I needed to hire. Yet here I was, staring at a blank page, suffering from writer's block. Then, I recalled a colleague telling me how he used ChatGPT when he had to do a similar task. At that time, ChatGPT was relatively new and all the rage, especially at the data analytics/web 3 technology company where I worked. I, however, approached it with a significant amount of skepticism. I refused to accept that some computer application could write something better than I could. I must also admit that I had no idea how to even use it and felt somewhat like a caveman staring at a cell phone trying to figure it out.

But after another wasted ten minutes, I gave in and downloaded it. I created an account and watched starry-eyed as the words populated on the previously blank page. It was like the streaming green code from *The Matrix*! I was fascinated. I had a draft. And it was good!

Valerie Capers Workman, Esq.

Now let's be clear—I had a *draft*. A starting point. It wasn't a finished product. The computer wasn't doing the work for me. The artificial intelligence that powered the app wasn't going to steal my job. It was a tool that allowed me to do my job…and better. We tend to think of the Industrial Revolution as a thing of the past, but it really never ended. With every generation, new technology and innovation continue to redefine how businesses operate and how we live our lives. From the full utilization of iron, steel, coal, steam, and electricity many years ago to the factory assembly line, telephone, computer, and internet, more recently, the human mind is constantly creating new ways to better our lives and maximize our full potential.

As a human resources executive, I am charged with helping my company attain its mission and generate revenue by maximizing its main resource: its employees. While I—and almost everyone—utilize some form of AI daily without giving it a second thought, the mere mention of artificial intelligence in the workplace is enough to stir up earnest feelings of anxiety and trepidation. Why? Why are we so reluctant to embrace this new technology as we would the newest version of a popular smartphone? Why do we fear being replaced and made obsolete? Why are we afraid to embrace artificial intelligence as a means of positive transformation and progress?

I was incredibly intrigued when my former boss and mentor, Valerie, told me she was writing a book about this topic. I first met Valerie when we both worked in the legal department at Tesla. When she left the legal department to become the head of HR for the company, I followed and became her head of HR compliance. Working for Valerie changed the entire trajectory of my career, and I am so happy that much of what she taught me over the years will now be shared with you here in *Quantum Progression*.

One of the most challenging projects Valerie and I tackled together involved data analytics, statistics, predictive technology, and advanced programs powered by artificial intelligence. Prior to that, I never saw myself as a tech-savvy person. However, I learned just how critical staying current with cutting-edge technology and AI was to advance my skills, experience, and, ultimately, my career. Many of the lifelong lessons I gained from Valerie are included in this book, and many are separate from any considerations of AI. Valerie demonstrated to me that my *career progression may not be linear.*

Indeed, I left my role as Deputy General Counsel and Director of HR Compliance at Tesla, the most valuable company in the world at the time, to become the manager of a four-person HR team at a small start-up and took a considerable pay cut. I did it because I knew I wanted to become a Chief People Officer ultimately. I *reverse-engineered my career path* and created one that eventually led to my role as a C-Suite executive at a larger tech company less than a year later. I also had to *detach from my job title* and let go of my hard-earned and prestigious identity as an attorney so I could pursue what I was passionate about—HR and professional development. Doing all this required massive amounts of *courage, emotional intelligence,* and *continuous excellence,* not to mention many *champions, allies, influencers,* and an *interdisciplinary team.*

Our collective and individual progression will require the integration of people *and* machines, not the replacement of people *by* machines. Using ChatGPT to speed up my completion of a job description allowed me to take time to listen to and mentor a new employee—something no computer application can do (yet). I would be remiss not to use every tool at my disposal to do my job and do it well—AI included. Resistance to this is not only futile but foolish.

Quantum Progression sets forth an accessible, detailed, and straightforward path for any and everyone committed to improving their professional lives and embracing the new age of artificial

Valerie Capers Workman, Esq.

intelligence in the workplace. To fully appreciate the teachings in this book, you must let go of fear and weariness and accept that change is occurring all around us. As the Nobel Prize winner Marie Curie once said, "Nothing in life is to be feared; it is only to be understood. Now is the time to understand more so that we may fear less."

Candice S. Petty, Esq.

Former SVP of People, Flipside Crypto, LLC.,
Former Deputy General Counsel and
Director of HR Compliance at Tesla

Ethical Considerations

I believe it's important to briefly discuss the ethical considerations that should be taken into account as companies decide how to leverage the benefits of Artificial intelligence (AI) in the workplace. If you are at a level in your career where you have a say in how AI tools are used, you have an obligation to keep these considerations in mind as you help your company decide which tools to use and how your company will use them.

The rapid integration of AI technologies into the workplace has led to new ethical challenges that demand careful navigation. As AI tools take on increasingly complex tasks, this new work paradigm raises concerns about equity, fairness, and accountability. Before discussing how to advance in your career in this new landscape, it's important for me to outline the ethical dilemmas associated with AI in the workplace and briefly discuss potential strategies to address these issues.

Valerie Capers Workman, Esq.

Bias and Fairness

One of the most significant ethical challenges related to AI is the potential for bias in decision-making. AI algorithms learn from historical data, which can inadvertently perpetuate existing biases present in the data. For example, AI algorithms can analyze a wealth of data about individual consumer behaviors, including purchase history, location, and browsing patterns. This information can be used to personalize the shopping experience, offering product recommendations more likely to appeal to each consumer. However, this same data can be misused to engage in unethical price discrimination if the data is used to charge different prices to different consumers or groups of consumers based on legally protected characteristics such as age, race, gender, sexual orientation, disability, or religion.

To tackle bias in AI tools, companies should implement thorough audits of algorithms, conduct regular testing to identify potential biases, and invest in diverse development teams. A diverse team will bring unique perspectives and experiences to the table, helping to uncover hidden biases and develop more inclusive AI solutions.

Autonomy and Accountability

As AI tools become more sophisticated, they can make critical decisions without human intervention. However, this autonomy raises questions about who is responsible for the outcomes of AI-driven decisions. Establishing clear lines of accountability is essential to avoid unintended consequences and legal liabilities. Companies must clearly define the roles and responsibilities of AI tools and the humans overseeing their actions. Additionally, transparency in decision-making processes and open communication with stakeholders can help build trust and ensure that AI-driven decisions are accountable and explainable.

Job Displacement and Workforce Transformation

AI's ability to automate tasks and replace human labor can lead to significant job displacement. While AI can streamline workflows, it can also create anxiety among employees who fear losing their jobs. Organizations must consider how to handle these workforce transitions thoughtfully.

To address potential job displacement, companies should focus on reskilling and upskilling employees to take on new roles that complement AI technologies. Emphasizing human-AI collaboration can help employees embrace the changes and see AI as a tool to enhance their career mobility and productivity rather than a threat to their jobs.

Promoting Ethical Leadership

Ethical leadership is critical to guide organizations in the ethical use of AI. Leaders should prioritize transparency, fairness, and inclusivity, making ethics a core component of their decision-making processes. Leaders should demonstrate a commitment to ethical practices in AI deployment. They should also foster a culture that encourages open discussions about ethical dilemmas.

Emphasizing Ethical AI Development

Companies should prioritize ethical considerations while developing and deploying AI tools by conducting thorough audits of algorithms to detect and mitigate biases. Incorporating ethics into the entire AI development lifecycle is crucial. This includes ethical considerations in data collection, algorithm design, testing, and ongoing monitoring of AI tools. Review teams or committees could also be established to evaluate AI projects from an ethical standpoint.

Valerie Capers Workman, Esq.

Encourage Equal Access and Inclusion

A diverse and inclusive workforce can help identify and address potential biases in AI tools. Encourage equal access at all levels of the organization, including AI development teams, to bring different perspectives to the table. Companies should actively promote equal access and inclusion in hiring practices and foster an environment where diverse voices are heard and valued. Equal access training can help employees recognize and address unconscious biases that may impact AI development and usage.

Educate Employees and Stakeholders

Companies should provide training on AI ethics to empower individuals to make informed decisions and identify potential ethical issues. Educational programs on AI ethics can equip employees with the knowledge and tools to navigate ethical dilemmas proactively. Additionally, clear communication about AI implementation, its benefits, and potential risks will ensure that employees are well-informed and engaged in the process.

Establishing Clear Ethical Guidelines

Companies should develop comprehensive ethical guidelines that define the boundaries for AI use in the workplace. These guidelines should address equity, fairness, accountability, and transparency while considering the company's values and goals. Ethical guidelines should be accessible to all employees and stakeholders, serving as a reference point for ethical decision-making. Regularly reviewing and updating these guidelines will ensure they remain relevant in the ever-evolving AI landscape.

Conclusion

AI technologies offer incredible workplace opportunities but also present new ethical challenges. Organizations must proactively navigate these dilemmas to ensure the responsible and ethical integration of AI into the workplace. By prioritizing fairness, transparency, and accountability, companies can **leverage** the **transformative** potential of AI while maintaining a respectful and inclusive work environment. Navigating AI-created ethical dilemmas requires constant vigilance, adaptability, and a commitment to upholding ethical principles as technology evolves. Embracing ethical AI practices will not only lead to successful AI implementations but also contribute to a more just and sustainable future for the workforce and society as a whole.

Introduction

Quantum Progression: My Career from There to Here

After 10 years of strategically pursuing and excelling in roles with increasing scope and responsibility, I submitted my application for the position of head of compliance at Tesla and won the competition for that job. However, while equally fruitful, the first decade of my career showed an apparent lack of rhyme and reason for my career decisions. I had a game plan and was executing it. On the other hand, to family and friends, I appeared to be drifting. Back then, it was still the norm to stay at the same company for 10 to 20 years, but I was only working two to three years in a role and then moving on. I did not know it then, but I was ahead of my time in understanding that loyalty to a company is misplaced energy. You owe a duty of excellence to the company where you are employed, but your loyalty belongs to yourself and your career. Understanding this difference is critical to maintaining momentum in your career advancement trajectory.

The first decade of my professional career began when I graduated from law school at 26 and scored a position as an attorney with a Wall Street law firm, earning very close to a six-figure

Valerie Capers Workman, Esq.

salary. My parents told anyone and everyone that their daughter was a Wall Street lawyer. I know I broke their hearts when I quit that job six months after I started, but I quickly realized the firm's culture was not a good fit for me. The attorneys in the firm were predominantly white and predominantly men. I found it extremely difficult to assimilate into that environment while simultaneously trying to learn how to do the actual work. I felt unprepared for Wall Street. I did not like that "fish out of water" feeling.

However, during my short time there, my career goals started to take shape. I noticed that despite the wealth and power the partners wielded at the law firm, they jumped whenever the clients called. The clients, although they were government officials who presumably had nowhere near the wealth of their lawyers, had the real power. The clients paid the bills. While my time at the firm was short, the learnings from that stint stayed with me. I learned I wanted to be on the side of the equation that made the decisions. I also learned I wanted to earn a generational wealth-building income equal to the lawyers' and bankers' compensation.

I also learned I wanted a career that focused on the greater good. The firm I worked for served as municipal bond counsel to U.S. cities seeking to raise capital for city projects. Those bonds were sold to fund essential community services like highways, bridges, schools, and recycling facilities. Before joining the firm, I thought doing good and earning good money were mutually exclusive. Based on the money the partners and bankers were earning, I learned I did not have to sacrifice wealth to work in jobs that made a difference in the world. So, in addition to calling the shots, I added a component of serving a greater good to the requirements for my dream career. While many of my jobs after the law firm did not always offer opportunities in altruism, the goal stayed with me.

When I resigned from the law firm, I had a general vision of my ultimate role—a lucrative senior executive job with a company that was doing good in the world. Note that this goal did not include a specific role, field, or industry. I was open to what that ultimate job would be and to take on roles that would help get me there. So, I started thinking about the core competencies needed to be qualified for that type of role. I decided a foundation built on marketing and communications would be most beneficial. Marketing because all jobs require the ability to sell thoughts, ideas, and strategies. Communications because the ability to communicate effectively—writing and public speaking—would serve me well in any role. (It turns out I was right.)

After a couple of exploratory early roles, I eventually hit my stride. The first signature role was the job I landed with Jazz at Lincoln Center (JALC), where Pulitzer-Prize Winning Composer Wynton Marsalis is the Artistic Director and Leader of the Lincoln Center Jazz Orchestra. One Sunday, I read in the *New York Times* that JALC was seeking someone to fill their Director of Marketing & Communications position. The role required experience with performing arts marketing, subscription ticket sales, and audience development. I had none of that. What I did have was a passion for jazz music gained from my father. I grew up listening to John Coltrane, Billie Holiday, and Miles Davis. When I applied for the job, I did three things that, once again, I instinctively knew I needed to do. I created a resumé formatted like a one-page newsletter, with various stories on my love of jazz and my marketing experience in other non-profit organizations. Then, using the Six Degrees of Separation theory (way before LinkedIn), I sleuthed out the connection I had to Lincoln Center and asked them if they would submit my resumé on my behalf. I scored an interview, crushed it, and got the job. I worked there for three very successful seasons. Great times.

It is important to note that another dynamic was at play when I landed that job. Three months after I started, my boss later shared

Valerie Capers Workman, Esq.

with me that he took a lot of heat for hiring me—from the JALC Board of Directors, the Lincoln Center leadership, and even some of the JALC execs. Remember, I had zero experience in performing arts marketing. What I have not yet shared with you is the fact that I was hired to promote JALC's first season as an independent constituent at Lincoln Center. JALC was preparing to join The Metropolitan Opera, the New York City Ballet, and the New York Philharmonic, to name a few of the venerable resident organizations on the world-renowned campus.

The first season had to be a resounding success, and here was my boss hiring someone who didn't know what a season subscription was and whose experience marketing concerts and events was gained as an undergraduate at Syracuse University. I, ever confident, was certain I could do the job. I believed then, as I do now, that skills and excellence are transferable, and it didn't matter that I lacked performing arts experience. To me, marketing was marketing. In addition, I had a passion for and a deep knowledge of the music. I knew I understood this audience. I knew I could do this job and do it well, and I knew I had an outstanding communications education I could lean on.

Still, it would have been demoralizing if I had known how everyone other than my boss felt about me getting the role. Thankfully, he didn't tell me about the heat he was getting right away. Three months into the job and about a week after I had scored a full-page cover ad on the *New York Daily News* Sunday edition touting JALC's inaugural season (*for free!*), my boss called me into his office. I had given him one of the two framed posters of the ad, and while holding it up, he told me, "They all said I shouldn't hire you. This changed everything. Great job, Valerie. Very cool!" He was visibly relieved and proud that he had made the right call. I did not realize until then that I was working under a microscope, but I showed my value and made my boss look good.

Quantum Progression

Just after my third year at JALC, I became pregnant with our first child. Leaving our house on Long Island at six o'clock in the morning, commuting on the Long Island Railroad 60 minutes into Manhattan, switching to the subway uptown, and staying late for concerts (even later for awesome after-parties) was exhausting. Eventually, I just couldn't keep up. So, I resigned from JALC and took a very low-profile marketing role with a firm on Long Island closer to home. Compared to the job at Lincoln Center, I was bored beyond measure. So, when I had our second son, I happily resigned to stay home for a few months to regroup. During this time at home, I refined my career plan and decided I was ready to put myself on the for-profit executive track, determined to be the Chief of something within the next 10 years.

When I was ready to return to work, I knew the move from non-profit to corporate would be difficult but not impossible. I knew the skills I gained and honed at Lincoln Center would be equally valuable in a corporate setting. Again, I looked for work that intrigued or interested me. When I read that KPMG was searching for a marketing manager for a newly created B2B team, I knew it would be the perfect role for me. I crushed the interviews and scored the job. At JALC, I learned I was excellent. At KPMG, I learned excellence is transferrable.

Knowing that excellence is transferable is one of my superpowers, and I'm thankful I discovered this early in my career. Not only has this knowledge fueled my confidence in my ability to take on new roles, but it has also helped me to spot undiscovered talent hiding in plain sight. I have built amazing teams that included exceptional performers who had been in the wrong roles or simply overlooked. I saw their potential and knew they could be stars in the right roles on my teams. My ability to spot talent—even when talent doesn't recognize themselves—is a skill that has consistently served me well.

Valerie Capers Workman, Esq.

At KPMG, I was hired as a marketing manager for a small, carefully curated team whose goal was to help the partners secure engagements with top-tier corporate clients and targets. After about six months on the job, my boss called me into his office and said, "Can I ask you a question?" I knew I was good at my job and had a great relationship with him, so I was more intrigued than concerned about his question. Before I could answer yes, he asked, "How do you get so much more work done than the rest of the team and at such a high level?"

Up until this point, I had not taken the time to assess my productivity compared to my teammates. I basically kept my head down and worked. My boss was looking at me expectantly, hoping I had some method I could share with the team, but my answer deflated the energy in the room. "I don't know," I replied. "I just work. But I will definitely think about that and let you know."

I knew I had disappointed us both with my response. That day, I walked out of his office determined to spend time developing an understanding of *why* I was exceptional in that role. I took inventory of the skills needed to excel and noted that those skills were not mentioned in the job requisition I responded to when I applied. Another learning to put in my back pocket—job descriptions rarely tell the whole story on which skills are *actually* required to succeed in that role.

I did so well at KPMG that I was promoted to Associate Director on a different marketing team within two years, which handled more traditional B2B marketing. While thrilled to be recognized as a rising talent, I did not enjoy the work. I couldn't re-engage the spark that made me a standout. However, I added yet another very important learning to the emerging development of my career progression formula. A former colleague of mine, a true rock star herself, used to always say, "All money ain't good money." I will paraphrase and say, "All promotions ain't good promotions."

24

Quantum Progression

Sometimes, you should say no but propose an alternative to show you are excited about moving up and contributing even more to the company. Frankly, you might need to say yes to keep your job. In that case, take the promotion but look elsewhere quickly before apathy ruins your hard-won reputation. Again, more on the potential pitfalls of progression later.

Shortly into my new role at KPMG, I knew it was time to make my next move back onto a legal career path. I just wasn't sure how to pair my marketing skills with my law degree. At the time, the legal profession was very cliquish. There were accepted norms that had to be obeyed. One such norm was not to leave the profession because it was extremely difficult to get back in once you were out. Deviate from that edict (i.e., step out of the profession or, worse, take a year or two off to have children), and it was virtually impossible to return to your former career path. I had been out of the legal profession for ten years when I decided to return. This was a challenge I knew I needed to solve to stay on track towards a Chief-level role. I was trying to find the right next opportunity but couldn't find anything that would work as a bridge back to legal work. Then came ENRON.

Working at KPMG during the ENRON collapse was like having a front-row seat to a trainwreck. ENRON is infamously known for self-inflicting one of the largest corporate bankruptcies and one of the most scandalous auditing failures in history. Its demise wiped out the hard-earned paychecks, retirement savings, pensions, stock, and equity of almost all of its 30,000 employees—virtually overnight. It was a corporate tragedy of Shakespearean proportions.

At the time of its demise, ENRON's accounting firm of record was Arthur Andersen. As a KPMG marketing team member, I knew all about Andersen. They were the world's top audit and advisory firm—if not by revenue, then certainly by reputation. Andersen represented some of the world's largest and most prestigious companies and directly competed with KPMG. For those readers who have never heard of the firm, Andersen was to the accounting industry what Apple and Google are to tech. Nine months after the collapse of ENRON, Andersen was found guilty of obstruction of justice for shredding documents related to their auditing of ENRON. Ironically, the convictions were ultimately overturned, but it was too late to save the firm. The audit issues and reports of document shredding would haunt the firm's reputation for years to come and lead to the firm's ultimate demise.

I witnessed firsthand the feeding frenzy that ensued as KPMG and the remaining "Big Four" audit and advisory firms PricewaterhouseCoopers (PwC), Ernst & Young, Deloitte, as well as high-end boutique firms with marquee clients like Mitchell Titus, competed with each other to recruit the excellent talent from Andersen. Most of Andersen's top partners and associates with excellent reputations and solid industry connections landed new jobs relatively quickly. It was my first exposure to high-stakes talent acquisition as everyone I knew at KPMG was getting calls from their friends and acquaintances at Andersen, and everyone at KMPG was suddenly a recruiter. The phones at KPMG were literally ringing off the hook. Another learning: be sure to keep your connections and contacts current.

As the biggest business story of the decade unfolded, I was riveted by the news reports. How did this happen? Tens of thousands of ENRON employees were out of work—their rock-solid retirement and investment portfolios evaporated. Watching the interviews of tearful former employees trying to figure out what they would do next was heartbreaking. As I watched these reports, I heard the phrases "failure of corporate governance" and "failure of compliance" repeatedly. Curious, I researched those phrases and discovered a new career path: Corporate Compliance. I learned that compliance was a combination of law, policies, procedures, and training designed to help ensure that companies follow the law and do the right thing. Essentially, compliance is about being a good corporate citizen—a responsibility ENRON apparently failed at miserably.

Learning about this profession was a source of inspiration. I found a career path designed to help ensure that a company behaves responsibly, protecting its employees, shareholders, and communities from foreseeable harm. I also learned that jobs in this field had the potential to reach pay levels in the high six figures. Soon after I set my sights on a career in compliance, KPMG posted my perfect next role—National Manager, HR Policies & Procedures. I went after that position with surgical precision. I knew my legal background, coupled with my marketing and communications skills, would be a perfect fit for the job even though I had zero experience working in Human Resources. So, I reached out to my internal connections and campaigned hard for the job. Perhaps a bit too hard. I scored and crushed the interviews, and I got the job. I was on the pathway back into law.

I later learned I almost didn't get the role because one of the HR leaders thought I was "too aggressive" in pursuing the position. At the time, I wondered if he would have characterized my passion for the job in the same way if I were a man. The team leader, who was also male, thought I would be great. Thankfully, his opinion mattered more. (Interestingly, the same pattern emerged years later when I applied for the Head of Compliance role at Tesla. One of the leaders thought I was "too aggressive" in my pursuit of the position, but the hiring manager fought for me.)

My time in HR at KPMG was like a tale of two cities. It was the best of times and the worst of times. I loved working with the HR Directors and Associate Directors across the country and using my legal skills to write HR policies and procedures, conduct training on HR processes, and update the employee handbook. My

27

Valerie Capers Workman, Esq.

marketing, communications, and legal background made me a perfect fit for the job, and I enjoyed the work. However, I soon found the job too easy; I did not feel challenged. After spending two years in the role, I started looking for my next position. Having succeeded in jobs that were national in scope, I knew I was ready to take on the responsibilities of being a vice president. I found my next perfect role at Wyndham Hotel Group ("WHG"), one of the world's largest hotel franchising companies. When I joined the company, WHG was part of a corporation called Cendant—a multinational conglomerate comprised of hotels and resorts, timeshare, car rental, and relocation companies, and they were preparing to spin off all divisions into separate entities. The role I succeeded in landing was Vice President, Franchise Compliance, WHG.

Winning the role at WHG was another illustration of the value of skills over functional experience. The role reported to the CFO. When I read the job description, I understood that the description placed a heavy emphasis on business process improvement and key performance indicators. While I had zero hospitality or franchising experience, I spent five years at KPMG reporting to and working with audit and advisory partners. I was fluent in CFO, and I understood that this CFO was searching for an executive to up-level the performance of the Franchise Administration team, agnostic of whether or not the executive had experience in the franchising or hospitality industries. As my first role at KPMG demonstrated, these skills came naturally to me. So much so that I didn't even know I had them until my boss pointed them out. I tailored my resume and cover letter accordingly. I got the interview, and we spent over 90 minutes discussing processes and ways to drive efficiency and productivity. The subject of the business of franchising itself was barely discussed. I got the job. From National Manager of HR Policies & Procedures to Vice President, Franchise Compliance.

Quantum Progression

At WHG, all of my functional skills came together synchronously. Here, I also learned I was a strong leader, a great mentor, and awesome at finding hidden and underappreciated talent. Finding talent comes easily to me. It's one of my superpowers. I simply look for people who remind me of myself when I was "coming up". I look for valuable and transferrable skills, a strong desire to get things done, a bit of an edge in having something to prove, a high EQ, and the ability to be highly collaborative. At WHG, I also expanded my skill set by becoming an expert in franchising, government relations and corporate compliance functions. My title change to Vice President Franchise Compliance & Government Relations reflected this career growth.

After six years working at WHG and almost three years as a Vice President, Franchise Compliance with WHG's former sister company, Realogy—owner of the most recognized names in real estate, including Century 21®, Sotheby's® and Coldwell Banker ®, I assessed my readiness for a Chief Compliance Officer or General Counsel role. While I was ready for those jobs "on paper", I decided I wanted to gain more comprehensive transactional experience. So, I started my search again and found the perfect role—Corporate Counsel Licensing & International at Focus Brands in Atlanta, Georgia. Focus Brands owns and franchises iconic food brands, including Cinnabon®, Auntie Anne's®, and Moe's Southwest Grill®. The company was looking for an expert in franchising, and I had the skills and expertise required for the role. In exchange, that job would allow me to strengthen my transactional skills, conduct more trademark maintenance work, and do more M&A (mergers & acquisitions) due diligence. This exchange was well worth the step backwards from a VP title. I got the job, crushed it, and the following year, I was promoted to VP, Legal. Sometimes, you need to move backward in your career in order to move forward.

In 2017, I was sitting in my glass-walled office in metro Atlanta in pre-pandemic full-throttled activity, and once again, I knew I could be doing more. This time, I was thinking back to my main goal of wanting to make a lasting contribution to the world. I began considering which company I could join to make a difference and, just as crucially, how to build equity ownership for my family. The only company that checked all the boxes for me was Tesla. I hit LinkedIn and saw that Tesla had recently posted my dream job—they were looking for someone to lead their global compliance function. Preparation meets opportunity. Rather than applying the traditional way, I sleuthed through LinkedIn, figured out who was the hiring manager, and contacted him directly. A couple of weeks later, I had a series of interviews, and a couple of months after that, I was moving my family (again) to follow my career. This time, we headed to Silicon Valley.

It is impossible to describe what it's like to work at Tesla, and I certainly won't try here. However, I can say my first two observations were startling. The first was recognizing the abundance of excellent talent at every company level. It was unlike anything I had ever experienced. Of course, there were standouts at every company I had worked for in the past, but at Tesla, standouts were everywhere—the norm rather than the exception. To be surrounded by this abundance of talent was invigorating. I was also surprised by the level of autonomy afforded to every employee. The freedom (actually, the mandate) to work on, contribute to, or even fix any problem was empowering. As an even greater surprise, I learned I had an appetite for adrenaline. I thrived in the relentless pace of the company. To constantly work at warp speed with an expectation of excellence in furtherance of the mission to save the planet was energizing.

At first, working at Tesla was a thrill ride. Within my first year there, I established a best-in-class framework for corporate compliance, and I played a key role in paving the way for the construction of the Gigafactory in Shanghai, China, by overseeing compliance from dirt to opening. At the same time, I saw opportunities to bring the same level of execution excellence to our employee programs. On one occasion, as the executive lead for Black at Tesla, I was particularly vocal at an Employee Resource Group planning session, speaking on ways Tesla could do better. Later, in December 2019, I received a text message—Elon wanted to see me. I was certain I was going to be fired. Still, I went into the meeting with a clear vision of the programs I thought we should introduce or enhance. At the end of the meeting, I walked out with a leadership role in HR, reporting directly to Elon.

Three months after moving from Legal to HR, I was leading the People function at Tesla during the most tumultuous time for corporate America in recent history. March 2020, COVID-19 hit the U.S. hard. The thrill ride became a life-or-death situation—literally. However, we were prepared. Tesla had established safety protocols developed for the Shanghai Gigafactory in 2019. By March, we were prepared with industry-leading safety protocols and tailored employee protection programs[1]. Then, in June, the murder of George Floyd at the hands of the Minneapolis police officers was filmed, and the gruesome and inhumane travesty ignited worldwide protests against police brutality. The horrific images impacted virtually every segment of society, and corporate America felt the resulting shockwaves. I was five months into my new role at the time of this global unrest. While the company did not make a public statement, I felt compelled to post a piece on LinkedIn titled "Bringing My Whole Self to Work at Tesla[2]". I described the challenges of navigating the dual roles of leading the

---

[1] Tesla's safety measures were published on the company blog on February 4, 2020 "Accelerating Tesla's Safety Culture" and on May 9, 2020 "Getting Back to Work."
[2] LinkedIn (Valerie Capers Workman, May 31, 2020)

Valerie Capers Workman, Esq.

People team during the pandemic and global unrest while being the wife of a Black man and the mother of Black sons. After the post went up, hundreds of workers worldwide, including Tesla employees, thanked me for sharing my authenticity and vulnerability.

Over the next two years, together with the stellar leadership team I assembled, we introduced a series of employee-focused programs designed to support the work life of Tesla employees. These programs, highlighted in the People sections of Tesla's 2020 and 2021 Impact Reports, included pay and equity ownership increases and enhanced physical and mental health benefits for all employees, hourly and salaried [3]. We published the company's first-ever Diversity, Equity, and Inclusion Report, replacing its *"Anti-Handbook Handbook"* with a comprehensive online employee guidebook. I also led the creation and rollout of Tesla's company-wide mandatory training, Respectful Recharge, which focused on the imperatives of treating each other with dignity and respect.

After four years at Tesla, I was ready to move on. I wanted to return to my legal roots and move to a General Counsel or Chief Legal Officer role with the right company. Positions at competing electric vehicle, battery, and clean energy companies did not appeal to me. I had worked for the most innovative company in these sectors. I was ready for a new industry but also wanted to continue making a difference in the world. I was open to discussing leadership roles with mission-driven companies. As I was contemplating my next move while at Tesla, I started hearing about a company called Handshake and was inspired by their business model—a democratizing opportunity for all students so that regardless of their zip code or school affiliation, they would have equal access to great jobs at great companies. When an executive legal search firm contacted me about the top legal spot at

---

[3] Tesla.com/impact

Quantum Progression

Handshake, I was already a fan. Meeting the CEO, executive team, and Employee Resource Group leads solidified my impression of the company, and I accepted the role as the founding Chief Legal Officer.

My exit from Tesla was a major news story in the tech world. I believed there was more work to be done, but I was confident it was time to make way for the next leader. In my statement to the media, I said that in leaving the company, I was handing the baton off in a better position than when I took the role and was proud of all I had accomplished at the company. I was equally proud to be called Tesla's "Top Diversity Advocate" — a label I wear proudly.

In my time between leaving Tesla and joining Handshake, I had the opportunity to reflect on my career and analyze the path I created for myself—the path that prepared me to help grow one of the world's most innovative and iconic companies from 50,000 to 100,000 employees. I began to understand that there was an art and a science to my career moves and the method I followed to advance.

I chose the term Quantum Progression to describe this method. Quantum has two meanings. In physics, it means a discrete quantity of energy proportional in magnitude to the frequency of the radiation it represents. In legal usage, Quantum means a required or allowed amount. In my method of career advancement, **Quantum means the career gains that you make are proportional to the energy you put in.** I chose the word *progression* to complete the term because the word, often misused, does not mean forward movement. **Progression means the process of developing or moving gradually towards a more advanced state**. In my method of career progression, a more advanced state means a conceptual and not necessarily an actual leveling up. In my career advancement equation, all movements—up, down, forward, backward, and lateral—can be positive.

The career moves I made ahead of my time are now just in time

Valerie Capers Workman, Esq.

for this new world of work in the dawning of AI. This career methodology is not for everyone. However, if you are comfortable with exerting sustained bursts of excellence, open to moving in whatever direction the right opportunity may take you, and committed to giving back as you move up, then I assure you Quantum Progression will take you wherever you want to go.

Ready?

THE ART: GET READY

"Only different in your mind. You must unlearn what you have learned."

~*Yoda*

Conquer Your Fear Of AI

"The positive power of AI will become clearer every day."[4]

~ Josh Bersin

Conquer Your Fear: The 1st Superpower Required to Advance Your Career

If you are concerned about Artificial Intelligence, or AI for short, then you are paying attention to the changing world around you. You know AI will transform how every job or task is done. Understandably, this knowledge can cause fear. Not the type of fear that impacts your daily life, but most likely, the kind that occasionally reminds you that you should be doing *something* to safeguard your future earning potential. This undercurrent of fear is just powerful enough to unnerve you when thoughts of your possible obsolescence surface from your subconscious to your waking moments. But being the bright, intelligent, and capable

---

[4] Why Is the World Afraid of AI? The Fears Are Unfounded, And Here's Why, Josh Bersin.Com, April 6, 2023.

Valerie Capers Workman, Esq.

person that you are, you know fear can be debilitating. After all, you have enough on your plate. There is no room to pile on something as amorphous and fanciful as fear of a future HAL 9000[5] or Skynet[6] takeover of humanity. Rather than confront them, you suppress those fears and focus on the realities of your day. Not quite in the *Ignorance Is Bliss* camp, you are more likely on team *Accept the Things You Cannot Change*. However, whether you acknowledge it or not, your fear of AI will persist, leaving you with two choices—accept and conquer the fear, or ignore it and concede control of your career to fate, luck, and chance.

The purpose of this chapter is to encourage you to acknowledge your fear of AI. By acknowledging this fear, you will be empowered to conquer it. To survive and thrive in the age of AI, particularly in pursuit of your career's (Quantum) progression, you must first proactively face the incursion of AI head-on. Rapid advances in AI are on the verge of transforming entire industries[7], redesigning job functions[8], and simultaneously presenting new challenges along with new opportunities[9]. While governments and businesses are

---

[5] HAL 9000 is the main "character" in the 1968 film *2001 A Space Odyssey.* HAL (Heuristically Programmed Algorithmic) computer is the artificially intelligent crew member who ultimately takes over the Discovery One spacecraft and kills the human crewmembers on board. It's a must-see film for sci-fi fans. The film directed by Stanley Kubrick is based on the novel by Arthur C. Clarke.
[6] According to the *Terminator* film series, Skynet is the defense system created by Cyberdyne Systems. It becomes self-aware at 2:14 a.m. Eastern time, August 29, 1997. In a panic, Cyberdyne tries to shut down Skynet, which takes over all of the global nuclear weapons and destroys most of the human race as it defends itself. It's a must-see film series for sci-fi fans. Directed by James Cameron, *Terminator* stared Arnold Schwarzenegger as the cyborg.
[7] *Will This Fruit-Picking Robot Transform Agriculture*, by Jyoti Madhusoodanan (The Guardian, May 28, 2022)
[8] *How AI Is Already Reshaping White-Collar Work,* (Wall Street Journal Video, July 6, 2023)
[9] The World Economic Forum's "The Future of Jobs Report 2020" predicts 85 million jobs globally will be replaced by AI by 2025. The same report also

38

moving to control the proliferation of AI[10], you should not rely on these protective measures alone.

In this chapter, I will discuss the importance of conquering your fear-driven response to the AI-driven changes that will shape your career prospects. My goal is to equip you with the courage you need to navigate and maintain control of your path to career advancement. Courage will serve as the foundation, or more accurately, as the well you will need to continuously draw from as you move further and further away from your comfort zones. When it comes to your career advancement, Quantum Progression requires constant movement. To make the moves you need to make at precisely the exact time you need to make them, you cannot let fear get in your way.

AI In Manufacturing: Factory Work as the Newest Tech Job

AI, in some form, can now be found in virtually all machines in the workplace. Wherever *a* something is required to *do* something, AI will be found. However, when it comes to intricate and complex movements, humans are still superior. One of the most notable demonstrations of the superiority of human dexterity can be found in Tesla's Gigafactory in Fremont, California. In 2018, when the company was tooling the facility to produce the highly anticipated and ultimately wildly successful Model 3 electric vehicle, Elon Musk intended to use a robotic workforce that he called the Alien Dreadnought to produce the cars. This plan was abandoned when it was found that humans were better at assembling cars with robots

---

indicates that AI can potentially generate 97 million new roles (Kalina Bryant, *How AI Will Impact the Next Generation Workforce,* Forbes, May 31, 2023).

[10] Pressured By Biden, AI Companies Agree to Guardrails on New Tools, by Michael D. Shear, Cecilia Kang and David E. Sanger, The New York Times, July 21, 2023.

playing their position as assistants.[11] Today, as machines continue to learn how to get better at the tasks they are designed to do, the relationship between person and machine is still symbiotic, with machines serving to increase human capabilities even as these machines continue to learn. As a result, every job involving an AI-powered device has the potential to become a grey-collar tech job that offers the opportunity to gain proficiency in AI skill sets. These newly created roles provide everyone willing to consider a career in manufacturing a means to secure an on-ramp to a lucrative career in tech—a compelling reason to embrace AI rather than fear it.

AI In All Human Operations: Faster and Smarter

Similar to the insurgence of AI in manufacturing, AI (particularly generative AI which can create "original" content such as texts and art) is also being used to enable humans to increase efficiency, effectiveness, and excellence in the performance of non-industrial roles. AI tools are now available to support virtually every job function. Like AI for manufacturing, AI tools for these roles will enhance, not supplant, the human workforce. And while AI today cannot yet compete with the common-sense reasoning of humans[12], these tools offer the opportunity for entry-level through C-Level workers to get work done faster and with increasingly greater degrees of excellence. AI provides the opportunity for skill expansion and professional advancement. The faster workers adapt to the integration of AI into their job function, the faster they will uplevel their performance and accelerate their career advancement.

---

[11] As issues mounted and delays became more prominent in the Model 3 ramp, Tesla and Elon Musk were forced to abandon the idea and instead adopted a manufacturing system that uses machines and people (Simon Alvarez, Teslarati, April 24, 2020).
[12] AI today is unbelievably intelligent and then shockingly stupid (Computer Scientist Yejin Choi, TED2023).

Gather Your Courage

"I learned that courage is not the absence of fear, but the triumph over it."

~ *Nelson Mandela*

Courage In Action: From Manager to Vice President

In my spare time, however infrequent that may be, I coach seasoned executives and mentor rising stars (and sometimes vice versa). In these settings, one of the most common questions I am asked is how I made the move from National Manager, Human Resources Policies and Procedures for KPMG (a Big Four audit and advisory firm) directly to Vice President of Franchise Compliance for Wyndham Hotel Group (one of the world's largest hotel franchisors). As an aside, I should note that THE most common question I get regarding my career is, *What was it like to report directly to Elon Musk?* I've never given the exact same response to this question, and I probably never will. Like all pivotal moments in life, distance and reflection continuously inform my perspective on that topic.

Valerie Capers Workman, Esq.

Going from National Manager to Vice President was a rare leap upwards on the corporate ladder, an entirely different job function, and a different industry—a Grand Canyon jump in career terms. This move put me on the path to a C-level role, and like the greatest all-time figure skater, Surya Bonaly[13], landing her record-breaking backflip in an Olympic competition, I stuck that landing by refusing to be told what goals I should and should not pursue. I did not let anyone tell me what was possible. I discussed this move in my *Introduction* to this book. I was fearless in my pursuit. I was confident in my ability to excel. And excel, I did. I crushed that job!

When you are confident, you move, speak, and act differently. Courage fosters confidence. However, it's important to note that confidence without the required skills and information to back it up can lead to career self-destruction. Few combinations are more disastrous in a career setting than being loud and wrong. Nonetheless, by empowering yourself with confidence, you will be able to create, pursue, and secure opportunities that fear would otherwise have caused you to miss. Courage is the foundation on which to build everything you need to thrive in your career, particularly as you navigate the challenges and opportunities of AI in the workforce. Without courage, you will make fear-based career choices, which will deter you from advancing in your chosen direction. Courage will enable you to control your career moves even when unexpected roadblocks appear. But can courage be taught? Can you actually learn to be courageous? Yes, and yes.

---

[13] Surya Bonaly, one of the greatest figure skaters of all time, placed second at the World Championships for three consecutive years (1993-1995) and is a five-time European champion. Bonaly is recognized as the only skater, male or female, to land a one-foot backflip (9 Reasons Why No One Compares to Figure Skater Surya Bonaly, by Joseph Erbentraut, Huffington Post, December 15, 2014).

42

Courage: The 2nd Superpower Required to Advance Your Career

Courage is the second superpower source you will need to advance your career. As an executive with three decades of leadership experience, I have had the opportunity to see a multitude of professionals make tremendous strides in their careers as a direct result of their ability to act with courage and seize or create opportunities for themselves. In this chapter, I will discuss the vital role that courage plays in the progression of your professional life. You will need the courage to move multi-dimensionally and navigate the pathways that AI will open and close throughout your career. Courage will push you to apply for and land roles that defy linear progression. Courage will give you the confidence to take on star-making projects that no one else wants to do. Courage will ensure you make career moves just in time instead of waiting until the time is right. Courage is the power skill you need to control your career destiny. With courage as your superpower source, you will be able to take advantage of opportunities, overcome obstacles, and achieve limitless career success.

The Quantum Progression Five Components of Courage:

1. Positive Self-Talk
2. Input Management
3. Worst-Case Scenario Planning
4. Best-Case Scenario Planning
5. Higher Power Thinking

Valerie Capers Workman, Esq.

Positive Self Talk

The most important conversations you have each day are the conversations you have with yourself. The choice as to whether those conversations are positive or negative is entirely up to you. However, it is important to understand that speaking positivity into your psyche is a building block of courage. The power of positive affirmations is well-documented.[14] The impact of positive self-talk on your professional (and personal) life cannot be overstated. Developing the habit of consciously replacing negative thoughts with optimistic and inspiring affirmations is a mission-critical skill that will enable you to advance in your career continuously. If you find generating or sustaining positive self-talk is too challenging, you should find support in therapy, clergy, coaching, or some combination or subset of all three.

Positive self-talk is speaking to and about yourself with optimism, encouragement, and self-compassion. It entails cultivating thoughts that build you up rather than tear you down. Unsurprisingly, it is the opposite of negative self-talk. Negative self-talk can present itself in various ways, including self-doubt, criticism, and doomsday scenario thinking. We all tend more towards negative self-talk. It's a learned behavior that can be extremely difficult to unlearn. Negativity can be so ubiquitous and insidious that we fail to recognize we are doing this to ourselves. It is vital to be hypervigilant of the tendencies to speak negatively to and about ourselves and replace those thoughts with more positive and, consequently, more constructive thoughts. There is a direct causal connection between thoughts and actions. Positive self-talk will propel you towards your career goals, while negative thinking will ground you in inertia by convincing you that you are not worthy of

---

[14] *The Happiness Advantage* by Shawn Achor is one of the most well-known books on the power of positivity.

44

achievement. Negative self-talk is debilitating.

If your self-talk leans negative, one way to counteract this is by treating yourself with the same kindness and understanding you show to the people you cherish most. Listen to yourself when you are encouraging or uplifting the people who lean on you for support in *their* times of need. Whatever you do for them and say to them, practice doing and saying the same to and for yourself, even if you don't initially believe your own positive thoughts. Positivity is a muscle that must be exercised in order to grow stronger. Over time, generating positive thoughts in precisely the moments you need them will become easier. It is important to note that positive self-talk does not mean treating yourself as if you are flawless. Instead, positive self-talk, when done correctly, means accepting your flaws and viewing them as opportunities for improvement and development rather than as reasons to feel unworthy of success.

Words have power, and the most powerful words are those you speak to yourself. Your ability to act with courage will depend on your ability to encourage yourself in the moments when courageous action is required. How will you know if you are saying the right things to yourself? If you begin to feel hopeful—even if only cautiously at first, you will know your thoughts are moving you in the right direction.

Input Management

Your ability to progress in your career depends on your ability to control the information you allow to permeate your psyche. Maintaining control over the information you ingest is essential to cultivating positivity and, by extension, courage. You must practice mindfulness regarding the social media you ingest, the people you associate with, and the ideas you allow yourself to ponder. The news media, social media, and toxic relationships are all sources of negativity that have the potential to seep into your psyche and alter your perspective. Over time, this negative input can become such a normal part of your daily existence that you begin to believe these negative and self-defeating thoughts are your own when, in actuality, they are the manifestation of negative input. Like the folklore of Dracula that warns it cannot enter a home without being invited in, these negative thoughts can't enter your mental space unless you open the door. And like Dracula, these thoughts are excruciatingly difficult to cast out once they get inside your head.

Most of the input that comes knocking on your psyche is outside of your control, making it all the more important to control what you can. Curating your input means actively seeking mentors, coworkers, friends, and family members who sincerely believe in your capabilities and ambitions. You must also curate your media consumption so that only positive, uplifting, and inspirational content is allowed on the feeds you view. Code your own algorithms so that you swipe left on any content that makes you feel envious, angry, frustrated, or depressed—or any emotion that makes you feel worse than before you viewed it.

Effectively curated input provides a supportive ecosystem that will promote your professional growth, assisting you in recognizing your strengths and addressing areas in which you may need to improve. The more you seek this positive input, the more you will find it. If any input in your life is causing you to feel bad about yourself, you may need professional help to understand why you allow this input to permeate your life and hold you back from pursuing success. If there are no voices in your life telling you that you deserve better, you are listening to the wrong voices. Change your input.

Worst-Case Scenario Planning

Fear of the way things *might* turn out has the potential to prevent us from embracing opportunities that would otherwise move us forward. We must ask ourselves, *What are the worst things that can happen if I take this risk?* In doing so, we ensure that our choices are based on accurate information. By investigating these possible stumbling blocks and acting out various hypothetical situations, we can gather vital insights that help us handle and reduce risks effectively. Being courageous does not mean acting irresponsibly. It means taking sensible chances in order to progress in your career. It will be easier to make well-informed choices that lead to career advancement when you consider the potential risks alongside the possible rewards.

Fear of the unknown can cause inertia, keeping us from exploring new chances. Conquering fear and uncertainty requires turning amorphous concerns into concrete scenarios we can game out so we can address them. You can take back the power your fears have to prevent you from taking risks by facing those fears head-on and challenging yourself to consider the possible outcomes of the worst-case scenarios. Making this exercise an ordinary part of your decision-making process will enable you to confront the unknown with courage and self-assurance, knowing that you have carefully considered every realistic and foreseeable possibility.

To conduct a realistic review of the risks involved, you must first ask yourself what the worst conceivable consequences could be. The time before you take action is when you are in the best possible position to make a dispassionate evaluation of the likelihood of undesirable outcomes and compare that probability to the potential advantages of taking the risk. This analysis offers a fair and objective perspective, which will help you make well-informed decisions. Playing out the worst-case situations enables you to uncover potential traps and weaknesses in your action plan. This awareness will allow you to determine ways to reduce those risks and create contingency plans to deal with the possible consequences of these risks. In addition to the tactical benefits,

Valerie Capers Workman, Esq.

worst-case scenario planning also provides emotional benefits. Identifying and planning for adverse contingencies can create a feeling of control, calming your fear of moving forward. Knowing you are prepared to recover from possible failures or setbacks fuels self-confidence and powers courage. Instead of diving headfirst into high-risk strategies, you will be taking calculated risks.

Best-Case Scenario Planning

Fear of success can be as debilitating as fear of failure. To overcome this fear, it is important to ask yourself, "What are the best things that can happen if I take this risk?" You need to actually say the words—out loud. If the sentence stirs uncomfortable emotions, you are likely afraid to succeed. This is a common fear because success means change, and fear of change is also very common. Similar to worst-case scenario planning, you will overcome the fear of success by visualizing the positive consequences and gaming out strategies to plan for the good things that may come your way as a result of taking calculated risks. If you don't do this exercise, you could suffer from a general feeling of anxiety that, if unidentified, will persist and prevent you from moving forward, though you will have no idea of the root cause of your inability to gain and maintain momentum.

The fear born from concern about moving forward in your career is valid. Progress will bring change. Your work schedule could be different, your duties and responsibilities could expand, and expectations regarding your performance will be elevated. These changes are just a few of the byproducts of achievement. Nevertheless, if you make a conscious effort to visualize and accept the benefits of taking a risk, you can adjust your perspective and open yourself up to the opportunities that success may bring. Planning for the best-case scenario allows you to shift focus from potential risks—often a byproduct of worst-case scenario planning–

–to the potential benefits of your plan. When you focus on the best possible outcomes and plan how you will adjust to your new situation, you will gain the confidence you need to overcome the nagging doubts about your ability to handle your new circumstances.

Higher Power Thinking

When courageous action is required, believing in something greater than yourself can be a source of inner strength and resilience. Belief in a higher power enables us to maintain healthy boundaries in our interactions with other people and helps us prioritize our most important values. It serves as a reminder that the opinions and validations of others do not completely define who we are. We can make decisions that align with our higher purpose when we remain rooted in our ideas and values rather than allowing ourselves to be persuaded by the expectations of others.

Belief in a higher power gives us the confidence to maintain control over the course of our lives and the ability to write our own narratives. The realization that we have the ability to make decisions and take action advances us forward in our efforts to realize our professional ambitions and triumph over any challenges we face. When we put our faith in a higher power, we become less dependent on the validation or acceptance of others. This lessens the anxiety that we will be judged by others, enabling us to make brave decisions without being immobilized by the thoughts and feelings of others. Knowing that our journey is being directed by something more than the opinions of others allows us the freedom to concentrate on our expansion and maturation.

Belief in a higher power also enables you to cultivate a sense of gratitude and humility—characteristics that are important to possess to ensure your courage does not foster a belief that your courage gives you the right to bully and oppress others.

Increase Your EQ[15]

"In a study of skills that distinguish star performers in every field from entry-level jobs to executive positions, the single most important factor was not IQ, advanced degrees, or technical experience—it was EQ."

~ *Daniel Goleman*

EQ In Action: Pitch Perfect Response to Office Drama

When I speak publicly about the time in my career when I needed my EQ to be at its zenith, most people think I am referring to my time at Tesla. My leadership role at Tesla indeed required me to draw on every facet of my executive leadership skills. However, the most challenging stretch of my career occurred years before I joined Tesla. I will speak in general so I don't identify the company or my role during this time. Suffice it to say I unexpectedly encountered a very challenging work-related situation. Out of nowhere, I found myself the target of the

---

[15] Emotional intelligence (or "EQ") is defined as the ability to understand and manage your own emotions, as well as recognize and influence the emotions of those around you. The term was first coined in 1990 by researchers John Mayer and Peter Salovey but was later popularized by psychologist Daniel Goleman, *Why Emotional Intelligence Is Important in Leadership*, Lauren Landry Harvard Business School Online, April 3, 2019.

typical office drama I had always avoided. Without warning, I had to work very hard to stay above the fray. My ability to manage my emotions during this time enabled me to think clearly and make all the right decisions when clear-headed action (or rather inaction) was required. When faced with this on-the-job adversity, I was able to preserve both my professional reputation and the career-critical relationships I maintain to this day. While the events were unfolding, I resisted the temptation to act in service of the emotions I was dealing with at the time. I could have indulged in actions that would have provided short-term emotional rewards. Instead, I controlled my emotions and thought about the benefits I wanted to reap from my hard work 5, 10, 15 years from those moments at that job.

To this day, I am thankful I had the tools to navigate a stressful and challenging situation by acting in furtherance of the long-term benefits of clear thinking. Colleagues who worked with me during this stretch still marvel at the way I handled the situation. Today, I have a good chuckle of relief when I think about how drama could have permanently derailed my career by possibly ensuring I never would have made it to Tesla. Ironically, it was due to this experience that I learned I could handle high-stress situations efficiently and effectively while performing with excellence. When the time came to use these skills at Tesla, I didn't question my ability to lead through challenging circumstances. I had already been tested. But can emotional intelligence be taught? Can you actually learn how to increase your EQ? Yes, and yes.

EQ: The 3rd Superpower Required to Advance Your Career

I have seen EQ deficiencies derail the careers of some highly talented individuals—hourly through executive. As seasoned as I am, watching someone implode a promising career due to a low EQ is always disheartening. To move multi-dimensionally and navigate the pathways that AI will open and close over the course of your career, you will need a high EQ. The irony is that if your EQ is low, you probably do not know this. And you can't cure a deficit you don't know exists. This is why it is critical you spend the money to have your EQ tested. A simple Google search will display a plethora of options. In terms of your career progression, paying for this test is the best money you will ever spend. Knowing your EQ is not optional. EQ empowers your leadership skills, enhances your ability to foster productive teamwork, and makes it easier for you to receive and act on feedback relative to your performance, all of which ultimately result in sustainable career progression. As AI continues to make its way into workforce tasks, a high EQ—which AI does not yet have—will essentially be the equivalent of career advancement insurance.

THE FOUR COMPONENTS OF EQ ELEVATION

1. Self-Awareness
2. Self-Management
3. Social Awareness
4. Relationship Management

Self-Awareness

Self-awareness—or the capacity to perceive and comprehend your feelings, skills, and shortcomings, as well as the effect you have on other people—is the foundation of EQ. Capacity to perceive is the key phrase you can't speed past. I have often given feedback to team members, and their first reaction was to refute what they were hearing rather than consider whether the feedback might be valid. This lack of self-awareness can derail the most talented professionals, even warm, friendly, and collegial people. Having a low EQ does not necessarily mean you are perceived to be rude or obnoxious. It means, among other detriments, that you are unaware of your impact on other people's emotions. If you actively work on increasing and cultivating your self-awareness, you will improve your decision-making, create stronger interpersonal interactions, and progress in your career.

The first step toward developing self-awareness is becoming conscious of your feelings as they occur. This entails locating the triggers that cause specific emotional responses, such as joy, worry, jealousy, anxiety, or anger. Individuals with a high EQ are able to manage their emotions more effectively when they become attuned to the triggers that cause those feelings, which leads to their ability to foster positive interactions and achieve better outcomes in the workplace—even during stressful workplace situations. Gaining self-awareness involves an honest evaluation of both your strengths and your limitations. When you are aware of your limitations, you are more likely to ask for and accept assistance. It continues to amaze me when high-performing individuals are reluctant to ask for help or assistance. This inability to collaborate almost always ensures that the project they are working on will not be as excellent as it could have been. People with low EQ believe that asking for help or engaging in true collaboration that values team members' input is a sign of weakness rather than a strength. If you are reluctant to accept help or input from others, you are limiting your opportunities to achieve performance excellence.

Self-awareness also improves your communication skills. Understanding the impact your communication style has on your coworkers will enable you to calibrate your cadence, tone, body language, and eye contact for maximum effectiveness and impact. This helps improve your connections with coworkers, which in turn contributes to building your reputation as an effective leader. Self-awareness is also helpful in recognizing the indicators of burnout, which is an important part of managing stress and building resilience.

Individuals with a high EQ are able to use coping methods and seek support to maintain their well-being.

Perhaps most importantly, individuals with a healthy sense of self-awareness accept ownership of their actions and the consequences of those acts for themselves and others, and they are more willing to apologize when appropriate. This level of accountability helps to cultivate trust and credibility, which leads to improved leadership qualities and unlimited prospects for career progress.

Self-Management

Self-management, also known as the capacity to effectively regulate and control one's emotions, impulses, and behaviors, is at the core of EQ. Self-awareness is useless if you don't have the capacity to self-manage. It requires becoming conscious of how your feelings influence your ideas and actions, as well as gaining the ability to control those feelings in a positive manner. Individuals can make more balanced and sensible decisions—even when they are under stress—if they are able to recognize when their emotions are running high and adopt healthy ways to channel them.

Work-induced stress is unavoidable, which is why it is important to practice stress management. Self-management involves adopting efficient stress management practices to prevent burnout and preserve general well-being. This is an essential part of self-management. Individuals can improve their ability to cope with stress and retain their attention and productivity by incorporating exercises like mindfulness and meditation into their daily routines, taking regular breaks from work, and seeking assistance rather than trying to do it alone.

Another benefit of self-management is the ability to resist impulsive emotions and think before acting. Before acting hastily, take a moment to pause and think about how the outcomes of your actions may affect those around you. The ability to regulate your impulses enables professionals to steer clear of needless disputes and make decisions that align with their long-term objectives.

Cultivating your ability to adjust to new circumstances is also essential to effective self-management. To advance in your career, it is necessary to possess the skill of adaptability, which includes keeping open to change, maintaining flexibility even in the face of uncertainty, and welcoming new challenges. Individuals who demonstrate the ability to flourish in various work situations and embrace change will have no problems keeping pace with AI.

Finally, the ability to accept feedback is a key component of self-management. Maintaining an openness to receiving feedback and incorporating feedback into your performance is critical. On-the-job promotions will accelerate advancements in your career if you actively seek feedback, demonstrate you can learn from your experiences, and continually strive to improve your performance.

Social Awareness

Social awareness, also known as the capacity to comprehend and empathize with other people's feelings, needs, and points of view, is one of the most important aspects of EQ. EQ can be developed in individuals through the cultivation of social awareness, which leads to improved communication, stronger relationships, and an overall increase in professional performance.

Active Listening: Developing social awareness starts with developing your listening skills. This requires not interrupting other people while they are speaking and keeping any thoughts about how to respond to their statements out of your head. Active listening will enable you to comprehend the message being communicated and allow you to respond with maximum efficacy and efficiency.

Interpreting Cues That Are Not Verbal: Facial expressions, body language, and tone of voice are examples of non-verbal cues that can provide insightful information about people's feelings and perspectives. Increasing your capacity to read and understand these signs will enable you to behave correctly and compassionately in a variety of situations that occur in the workplace. People with high EQ know how to read the room and successfully calibrate their responses accordingly.

Fostering Empathy: The ability to empathize with others is essential to developing social awareness. It requires putting yourself in the position of the other person, gaining a knowledge of their emotions and points of view, and demonstrating genuine care and concern. Through cultivating empathy, you will demonstrate your ability to foster a work environment that is welcoming, inclusive, and productive.

Recognizing the Cultural Differences: In today's increasingly varied workplaces, it is crucial for efficient communication and collaboration to acknowledge and appreciate the cultural differences among coworkers. Being socially aware enables you to navigate cultural differences in a sensitive manner and steer clear of misconceptions that can derail your career[16].

Relationship Management

Relationship management is one of the most important aspects of EQ. This skill entails conducting interpersonal interactions productively and establishing solid ties with coworkers. Your EQ can be strengthened by cultivating relationship management skills, which leads to your ability to foster effective teamwork, demonstrate expertise in conflict resolution, and uplevel your overall performance.

Establishing trust and rapport with others is the first step in managing relationships. Building trust is essential to maintaining healthy relationships. This can be accomplished by acting in a

---

[16] More on this topic later in Pitfalls: Unconscious Bias

reliable and consistent manner. You can cultivate an atmosphere of trust in which open communication and collaboration are encouraged by acting in a way that is honest, transparent, and accountable in your dealings with your coworkers.

Relationship management at work can be challenging. Disputes are an inevitable part of the dynamics of the workplace. Effective relationship management requires the resolution of disputes through the application of emotional intelligence. You will be able to find solutions to disagreements that are acceptable to both parties and stop misunderstandings from growing into full-blown arguments if you approach moments of conflict calmly and with respect. Another critical component of effective relationship management is the ability to provide feedback in a constructive and encouraging way. Professionals skilled at relationship management are also adept at motivating and inspiring their coworkers. They are aware of the capabilities and goals of the individuals in their work circles, and they adapt their behavior to demonstrate support and encouragement in moments when rallying team members is critical to a successful outcome.

Effective relationship management also requires efficient collaboration with coworkers who come from various experiences and have a wide range of skill sets. This skill will enable you to create the synergy that fosters innovation and problem-solving by recognizing the value of the contributions made by your coworkers regardless of their level in the organization—or yours.

Detach From Your Job Title

"Who you are is not what you do. If you get this twisted, when your title disappears, so does your self-image and worth."
~ *Christy Rutherford*

Title Detachment in Action: Moving Backwards to Move Forward

To clear all pathways to your career advancement, you must detach yourself from the appeal of job titles and embrace the importance of accumulating mission-critical skills and experiences. I do believe that job titles are important in that a title should clearly define a person's role and level in the company. However, it is important to know that titles are temporary and situational and have zero correlation to your value as a human being. I have firsthand experience with the transformational potential of detaching yourself from your title in order to prioritize skill development over cachet. At one point in my career, I made the conscious choice to leave a vice president role to take a director-level role in another company (and move my family to a new state).

Valerie Capers Workman, Esq.

The director-level job gave me the opportunity to provide expertise to my new company in the key areas of the role while gaining valuable experience in the areas of my background that I felt I needed to strengthen. In just over one year, I was promoted to vice president. The skills I added were critical to my growth to C-level readiness.

My career has fluctuated between pursuing titles for career-level progression and pursuing skills for career bandwidth. I encourage you to emphasize skill development and expansion over job titles in the pursuit of career advancement. Knowing that I was not my title made the optics of these moves easier for me to ignore. It is essential to develop this mental and emotional security because your professional and family circles might lament that you are "getting demoted," "moving backward," or "floundering". I routinely laughed at these assessments of my career without feeling the need to explain myself because I knew I had a game plan that I was executing perfectly. I was collecting skills and experiences like baseball cards. Possessing this wide range of capabilities has made me a better people manager and a more effective leader. To put a new spin on an age-old saying, I believe those who *have done* (rather than can't do) can teach with credibility. Detaching yourself from your job title is the key to ensuring your continued career progression in the age of AI.

Title Detachment: The 4th Superpower Required to Advance Your Career

In the workforce, an extremely high level of symbolic importance is attached to job titles. This is understandable. Titles celebrate and broadcast an individual's achievements and standing within an organization. On the other hand, titles do not always provide a reliable indication of an individual's actual talents or value to the company. Throughout your career, the attraction of prominent job titles may obscure your judgment when choosing your next role or accepting a promotion. Title envy can cause you to overlook rare and timely opportunities to develop new skills. It is mission-critical for you to understand that in this age of AI, skills are becoming more important than experience. Detach yourself from the reliance on job titles as a means of validating your worth and expertise. As you progress in your career, the lofty titles will come, and you will unquestionably deserve the prestige, salary, and accolades that accompany each of your accomplishments. However, acquiring essential skills should be your top priority,[17] especially if you are early or midway along your career advancement journey.

THE QUANTUM PROGRESSION: FOUR COMPONENTS OF TITLE DETACHMENT

1. Pursue Critical Skills Over Lofty Titles
2. Do Not Obsess Over Your Title
3. Focus on Your True Identity
4. Separate Work Life from Real Life

Pursue Critical Skills Over Lofty Titles

Your versatility and adaptability—key traits in the age of AI—increase in direct proportion to the range of your skills. Having a wide range of skills makes you useful to many employers. Rather than pigeonholing yourself into a certain job title, if you focus on skill development, you can easily transition across numerous job types and industries, expanding your career prospects. Although a prestigious job title may look tempting on a job posting, consider whether the role will be additive to your current skills. For example,

---

[17] Focus on Your Skills, Not Job Title: Get Hired by LinkedIn News, October 19, 2022

Valerie Capers Workman, Esq.

when weighing whether to go for the "Manager of Everything" job versus the position of "Director of Some Things," consider which role will be most beneficial to you over the course of your career. Moreover, if the manager role pays more than the director role, would you choose the latter because the title sounds better at conferences and family gatherings? Suppose the manager position is a "stretch role" that will allow you to immediately contribute to the company while gaining at least one critical new skill. In that case, the manager role is probably the better career move. Choose the manager role. The more impressive title will have to wait.

This is how you open multiple paths along your career progression. In addition to gaining skills on the job, continuous education is necessary for anyone interested in expanding their skill sets. By placing a high priority on skill development, you will perpetually be in growth mode. This dedication to lifelong learning encourages personal and professional development, leading to exponentially greater potential opportunities.

Do Not Obsess Over Your Title

All too often, the titles we hold at work frequently become entangled with who we are. We begin our introductions by stating our occupations, using these jobs to validate ourselves and establish our sense of self-worth. However, it is crucial to acknowledge that genuine identity goes well beyond the labels ascribed to us based on the occupations we have chosen. You will build a greater sense of purpose and satisfaction in your work life by knowing the hazards of over-identifying with your job titles and putting into practice tactics for sustainable self-worth impervious to the fluctuations of work situations and the impermanence of titles.

When you tie your sense of self-worth entirely to the titles you hold at work, you leave yourself open to being validated by others. This happens when you limit your sense of worth to your job. When you allow external criteria such as promotions, compensation, or hierarchical rank to determine your feeling of value, you are ensuring that your self-worth is conditional and situational. This over-dependence on other people's opinions can result in an unstable sense of self and derail your personal and professional development and fulfillment. When you identify too strongly with your job, losing sight of your most basic and self-defining values can be easy. Overvaluing titles puts your authenticity and well-being at risk. Overidentifying with your job title can lead you perilously towards unethical choices that might jeopardize your career.

Focus on Your True Identity

Setting aside time each day to focus on the things most important to you beyond the confines of your job title, such as your values and life goals, is essential. Determine the pursuits, causes, or hobbies that interest and excite you and are congruent with your core principles. Take the time to engage in some introspection to ground yourself in your identity and understand how your chosen line of work does not define your true self. Participate in volunteer work that brings you joy and personal satisfaction. Your professional and personal life will be more balanced and fulfilling if you focus on developing different aspects of your identity outside of your career.

Separate Work Life from Real Life

Work life is not real life. It is a fictional construct—a temporary moment that can change *at* any moment. You can quit or lose your job at any time. The company you work for was likely created before you started there. In other words, it does not exist because of your mere presence. Your work life can change in an instant depending on who is leading the company and who your team members are. The dynamics of your daily work life can change simply by the ebb and flow of employees who join and leave the company. The longer people stay at one job, the more they believe in the permanence of this construct, which is why it can be so devastating to leave a company after years of dedicated service— no matter the circumstances of your departure. It is important to understand that your job is not your real life. It is a *quid pro quo*[18] exchange where you provide a service, and in return, you receive the benefits you bargained for when you accepted the job offer. You should stay as long as you get what you need and want from your job. When the role no longer serves you, get a new position in the company or look for a new job elsewhere. Your real life is yourself, your family, your friends, and your community. By creating distinct boundaries between your job and personal life, you won't ever succumb to the false belief that you are your job title.

---

[18] Latin for "something for something"

Live Where You Matter

"Live Where You Are Celebrated, Not Tolerated."
~ *Original Author Unknown*

Living Where I Matter in Action: The Empowerment of Community Support

At one point in my career, I lived and worked in the same New Jersey town. At the time, this was the pinnacle of success in terms of work-life balance because it was the first time in my career that I did not have a two-hour-each-way, tortuous commute. The worst commute my husband and I had to endure was the commute we suffered through early in our marriage. To get to work, we drove from our home in Wyandanch, Long Island, to the Staten Island babysitter, to Sylvia's Restaurant in Harlem for breakfast, then dropped my husband off at his job in Long Island City, Queens along the way to my job in Montvale, New Jersey. Three hours each way. Good times. (Though I must admit the daily breakfasts at Sylvia's were a high point.) As the wife to a hard-working husband with his own professional career and mother to three young and active sons, living and working in the same town without having to compromise on my salary was a lifestyle achieved through equal parts hard work and perseverance. This was also the same five-year stretch when my husband worked year-round in Washington, DC, Mondays through Fridays. Our life/work situation meant that the physical and mental stress of day-to-day parenting was on me—school drop off and pick up, after-school activities, schoolwork, and co-managing our family-owned track club—all while I was vice-president of a global company. This was pre-COVID. While his job could have been 100% remote, that option was not available then. How our marriage survived those five years is a testament to the fact that miracles still happen. One thing that kept us going was my husband being home every weekend of those years. Admittedly, it was hard to convince our neighbors, families, and children that we were not divorced or separated. During the recession and rampant layoffs, we did what needed to be done to maintain our family income. Talk about the need for a high EQ. God bless our military families who live this life while also risking the ultimate sacrifice. Bless you all.

During this time, we were blessed to live in a community where working parents supported each other, there were tons of activities for the kids, and the schools were solid. I would be remiss in telling this part of my story if I did not note that we encountered the well-documented challenges of raising Black sons in a predominantly white suburban school system. However, the impact of this setting on our sons' educational and social experiences is a subject for a very different book. For the purposes of *this* book, I must point out it was due to the support of the parents—primarily the moms—of all races and backgrounds in that town that I was able to excel in my career. Living in this ecosystem was the key to my ability to push the accelerator pedal to the floor during the stretch of my career when I needed to be hyper-focused on excelling at work.

As you evaluate the moves necessary to advance in your career, it can be easy to overlook the role your address will play in your ability to deliver excellent results consistently at work. Is the commute going to be sustainable? Do you have access to community support systems? Is a work-from-home role compatible with your home life? These questions cannot be afterthoughts. Your community support system is as important as your support system at work. While you may not be at a stage in your life where you have complete control over where you live, it is important to control as much as you can. Where you live will have a major impact on your career velocity and trajectory.

Live Where You Matter: The 5ᵗʰ Superpower Required to Advance Your Career

The ability to plug into a supportive community is critical to your ability to progress in your career. Living in a supportive community will provide you with a strong feeling of belonging and the emotional support you need to live your best life. Do not rely solely on your work to fulfill your emotional needs. Depending too heavily on your work environment for fulfillment will cause you to make poor career decisions, such as staying too long in one role or doing the wrong thing when faced with an ethical dilemma. Yes, work should provide a safe and satisfying environment that allows you to perform to the best of your abilities. However, work is not and should never be a substitute for community.

THE QUANTUM PROGRESSION FOUR BENEFITS OF LIVING WHERE YOU MATTER

1. A Strong Support System
2. A Sense of Belonging
3. Reduced Levels of Stress and Anxiety
4. Maintenance of a Healthy Work-Life Balance

Valerie Capers Workman, Esq.

A Strong Support System

When striving for a successful and satisfying career, it is common to overlook the impact your living environment has on you. Your mental health and ability to advance professionally can be strongly impacted by the neighborhood or town where you decide to make your home. Living in a supportive community that accepts you for who you are will provide you with an environment that enables you to flourish in your career. The emotional baggage that comes with the experience of living in a community where you feel like an outsider can derail your career advancement.

An emotionally supportive community is there for its members to lean on in times of success and failure. This emotional support network functions as a buffer against the isolating effects of career challenges and enables professionals to navigate through obstacles with greater resilience than they otherwise would have been able to muster. This readiness to help one another cultivates a culture of empathy and compassion, which improves emotional well-being and positively impacts your life and career. Living where you can connect with people who will support you creates the perfect incubator for your professional growth.

A Sense of Belonging

You are more likely to experience a sense of belonging when you live in a community that accepts you for who you are, values your presence, and makes you feel like you are a part of something bigger than yourself. This sense of belonging has a significant impact on your emotional well-being as well as your self-confidence and overall mental health. You are more likely to feel a higher sense of purpose and fulfillment in your personal and professional life when you have a genuine connection with your community. This sense of belonging can serve as a safety net during times of career uncertainty and lower your anxiety when taking career risks for your advancement.

Reduced Levels of Stress and Anxiety

Being made to feel like an outsider in your community will result in higher levels of tension and anxiety, negatively impacting your professional performance. On the other hand, being a part of a community that provides support and encouragement will promote your mental and emotional health, provide a sense of safety, and significantly reduce stress. Because of this reduction in stress, you can focus on advancing in your career rather than wrestling with the emotional distress of feeling like an outsider. Reduced anxiety contributes to peace of mind, which fosters a feeling of serenity and contentment and provides a firm foundation for a high EQ.

Maintenance of a Healthy Work-Life Balance

You are more likely to be active in a community where you feel supported. When you participate in community activities, you will establish a good balance between the activities you do for work and those you do strictly for fun or personal satisfaction. Increasing your involvement in your community will decrease your reliance on your job as the source of all your joy, satisfaction, and value as a human being.

Community involvement and leisure activities are frequently deprioritized at work. Even in a post-COVID environment, employees are reluctant to discuss their interests outside of work or will apologize for taking mental health Paid Time Off. For this reason, it is important to work at a company—and, more importantly, report to a manager—who supports your frequent participation in community activities, not just your grand once-per-year vacation plans. I urge you to put equal effort and focus on your community life. Doing so will help you make bold bets and take calculated risks that will advance your career.

Manage Your Brand

"You ARE a Brand."

~ Nick F. Nelson

My Brand in Action: Maintaining My Authenticity in the Shadow of Elon Musk

When I applied for the compliance role at Tesla, I was sitting in my pre-COVID, glass-walled office in Atlanta, almost three years into a great job for a great company. I was the Vice President, Legal for Focus Brands, the owner of Cinnabon, Auntie Anne's, Jamba, etc. I led global anti-corruption compliance, negotiated international supply chain and global channels retail licensing deals, and maintained and protected the company's global trademarks. By virtually all measurement standards, I had officially "made it" —great title, great salary, great company, great neighborhood, great house. Still, I couldn't shake the feeling that I should've been doing more to make a difference in the world. As I thought about which company I would most like

Valerie Capers Workman, Esq.

to join, the only company that met all of my requirements was Tesla. I loved the company's focus on the realities of climate change and its conviction that electric vehicles could play a major role in saving our planet.

When I applied to work for Tesla in 2017, Tesla was well-known in corporate circles as a company where the senior executives maintained a low profile. While conducting my research on Tesla as I prepared to send over my résumé, I could not find the ubiquitous list of executives on the company's website. That was fine with me. I wasn't looking to join Tesla to become famous; my objectives were to help the company save the planet through its mission to accelerate the world's transition to sustainable energy, to continue to build wealth through equity ownership, and to continue to advance in my career. Four months after spotting my dream job on LinkedIn—Associate General Counsel, Compliance for Tesla, I was moving my family (again), this time across the country to California.

When I joined Tesla in 2018, I fully intended to be promoted to a C-level within two years. If you've come this far in this book, you know I believe that movement in any direction can ultimately lead to career progress. However, sometimes a step up *is* the best next move. My objective when I joined the company as the functional head of compliance was to be named the company's Chief Compliance Officer, a step up. The role required C-Level work, and by doing the job with excellence, I planned to earn that C-Level title.

By the time I joined Tesla, I knew exactly who I was as an executive and human being. I believed communication and transparency were the best ways to lead, especially during a crisis. I believed in the principles of Diversity, Equity, and Inclusion[19]. I

---

[19] Under my leadership, Tesla published its first-ever *DEI Report* (December 2020)

also knew there would be times during my tenure with the company when I would be seen first and foremost as the female Tesla executive or the Black Tesla executive (or both). I was already one hundred percent comfortable embracing and representing those facets of myself as a leader. Before joining Tesla, I had been "the only" many times in my professional life. I also knew my faith and my family—my real life—were my foundation. Most importantly, I knew if I ever needed to decide between my integrity and a hefty paycheck, fried bologna sandwiches would be fine. This is my brand. These are the values I walked in each and every day at Tesla.

You can see my values reflected in the email message I sent to the employees after the horrific murder of George Floyd.[20] You can see my values reflected in the *actual* work that Tesla put into protecting employees during the COVID pandemic (compared to the public drama surrounding the shutdown and re-opening played out in the media headlines).[21] You will also see my values reflected in the statement I made on behalf of the company during very public litigation[22]. I believe I am the only executive at a Fortune 500 company to take on the use of the N-word and other derogatory

---

[20] I was one of the first Black executives to publicly discuss what it was like to be in a corporate leadership position during the Summer of Protests. *Bringing My Whole Self to Work at Tesla* (LinkedIn Article, May 2020)

[21] *Getting Back to Work* (Tesla.com May 9, 2020) and *Health & Safety at Tesla* (Tesla.com July 2, 2020) were published by the company's VP of Environmental Health and Safety. These "playbooks" served as the models for the safe re-openings of companies in all industries across the country—a fact often lost in the headlines at the time.

[22] "While we strongly believe that these facts don't justify the verdict reached by the jury in San Francisco, we do recognize that in 2015 and 2016, we were not perfect. We're still not perfect. But we have come a long way from five years ago. We continue to grow and improve in how we address employee concerns. **Occasionally, we'll get it wrong, and when that happens, we should be held accountable**." (emphasis added) *Regarding Today's Jury Verdict* (Tesla.com October 4, 2021)

slurs at work, publicly and unapologetically.[23] You will also see my values reflected in the worldwide employee-focused programs Tesla rolled out during my tenure.[24]

Your "professional brand" is the impression that others have of you due to their interactions with you, your reputation and your communications in your work spaces, your industry, and on social media. This impression can be positive or negative. Your brand is how you present yourself to the world and includes the values you publicize and the values you uphold (which may or may not be identical). Whether you like this concept or not, you must understand that you are a brand. The question is whether you are taking charge of your brand or leaving your brand to the whims of external influences.

While the focus on who you are as a professional may not be as public as my tenure at Tesla, it is important to know that if you have a job (or plan to get a job), there is an audience watching what you say, what you do, and how you present yourself to the world. Whether you are early in your career or at the height of your C-level tenure, your career prospects in the age of AI will primarily depend on your professional brand.

---

[23] "…It can be empowering to take what was meant to harm you and instead redefine and own it on your own terms. This is by no means limited to one group of people—every community has ways of communicating to and about each other that, if used by someone not in that community, would cause pain and offense. This is a complicated social issue. But, at Tesla, it's a distraction away from our mission to try to debate the acceptable and prohibited uses of these words. So, we don't. Tesla expressly forbids all such slurs, epithets, or derogatory expressions based on any characteristics a person may have. Regardless of intent…"*Back to the Office: Excellence and Respect*, (Tesla.com, July 21, 2021)

[24] A relatively comprehensive summary of those programs is provided in the *Tesla Impact Report 2021* (pp.21-44, Tesla.com) and the *Tesla Impact Report 2020* (pp. 54-76, Tesla.com). Important to note that I did not accomplish these wins on my own. I assembled and led a stellar team of professionals.

Managing Your Brand: The 6th Superpower Required to Advance Your Career

Developing, guarding, and consistently cultivating your professional brand is essential for you to advance in your career and achieve sustainable success. A powerful brand allows you to stand out in a congested market, gives you access to previously unavailable opportunities, and encourages the kinds of relationships and partnerships that benefit you. You can develop a compelling brand that helps move your career forward if you define your values, manage your online reputation, nurture professional relationships through mutually beneficial connections, and remain loyal to your authentic self.

The Quantum Progression Six Components of Successful Brand Management

1. Decouple Your Professional and Personal Brands
2. Cultivate Your Work Identity
3. Foster Brand Consistency
4. Position Yourself as a Thought Leader
5. Maintain Your Authenticity
6. Curate Your Social Media

Decouple Your Professional and Personal Brands

Your professional brand is the collective representation of your abilities, expertise, values, and other distinctive attributes that set you apart in the workforce. It is important to remember that your professional brand is a reflection of you as a professional individual. It is the total of your reputation, the impression you provide to others, and the narrative that, taken together, forms your professional identity. A powerful professional brand establishes

Valerie Capers Workman, Esq.

you as an authority in your industry, enhances your credibility, and sets you apart from your peers. A compelling brand can pave the way for new professional opportunities, promotions, and even opportunities to start your own business if that's your ultimate goal. But here is a word of caution. You must create a distinction between your professional brand and you as a person. All too often, I see talented professionals oversharing their personal lives in the purported interest of authenticity or bringing their whole selves to work. We need to refine these terms to say that you need to be your authentic "work" self and bring your whole "work" self to work. I'll explain.

Some parts of who you are should only be shared with family and friends in your personal life. This does not mean you should hide any of those things, but you must carefully select (or rather curate) the parts of your life you want to highlight in furtherance of your professional objectives. For example, I can use some tried and true curse words when I am very happy or upset. (Oftentimes, the same words suffice in either situation.) But you will never see me use curse words in my social media posts. Does this mean I am pretending to be someone who does not curse? No. It means the part of myself that uses strong language is not the part I choose to present to the general public. Another great example? I am a proud Black woman who proudly wears her natural hair. I also have very strong feelings about the composition of chemicals that I previously used to straighten my hair[25]. But you will never see me post online content that rants about the chemicals in hair straighteners. What does that subject have to do with my expertise in Corporate Governance? Absolutely nothing. Am I less of my whole self if I choose not to discuss chemical hair relaxers in a public forum? I

---

[25] For the record, I have no problems whatsoever with the process of straightening Black hair. It is the lack of scientific progress with the chemicals themselves that I object to.

think not. One more example? I love to sing. I am proud of my voice. My love of singing makes me…well, me. If there is a karaoke night-out work function, I will attend. I am a team player who leads by example, so I will sing. I am also a huge hip-hop fan. Would I rap to a Missy Elliott track at a work function? Definitely not. Most of Missy's lyrics are not in line with my professional brand. I also love Whitney Houston, so I'd sing one of her songs. Work event? Whitney. House party with friends and family? Missy. My authentic self will show up in both places. One is my work self; the other is my personal self.

Some very talented professionals believe that having an "authentic" brand means sharing every thought, belief, and emotion they have without curation. They are causing severe damage to their otherwise promising brands. If this is a conscious choice, then so be it. If those stream-of-consciousness posts are intentionally on-brand and designed to increase content reaction percentages, then go for it. However, if you are not devoting time to the careful curation of the online presence of your professional brand, sharing only those parts of you that you want to highlight in furtherance of your career, then you are not managing your brand. You're just posting random content. It is important to know the difference and choose the path that works for you accordingly.

If your online content is not getting you the results you are striving for, you should consider a brand makeover. A wealth of brand experts are ready and waiting to help you align your brand with your career goals. And please be encouraged. No matter how far you have strayed from the professional brand you want to present to the world, a brand expert can get you back on track. (Just be prepared to delete some posts you worked hard to create.)

Valerie Capers Workman, Esq.

Cultivate Your Work Identity

Engage in some self-reflection to determine your primary strengths, distinctive abilities, and the value you bring to the table. Defining your value proposition is another crucial step. Define your one-of-a-kind value proposition, which consists of the skills, experience, and personal attributes you possess that set you apart from others in your field. Establishing a captivating brand starts with articulating your value proposition in a clear and concise manner. Establishing a powerful online presence is essential to creating a successful professional brand for yourself. Build and maintain a professional social media presence highlighting your knowledge, achievements, and thought leadership. Establish a strong presence on professional networking platforms where you can share your thoughts, interact with your colleagues, and expand your professional network. Note that you can create your professional brand at any stage of your career, even if you have yet to enter the workforce or are new to the workforce. Tenure is not relevant. You just need a clear voice, a fresh perspective, and a compelling point of view.

Foster Brand Consistency

Be sure to maintain coherence in the manner in which you communicate your brand across all of the different channels where you post your content. Ensure that your messaging resonates with your beliefs and unique value proposition. Trust is built via consistency, which also helps to reinforce your professional brand. Monitoring and actively managing your internet reputation is an essential aspect of managing your professional brand. You should perform regular audits of your digital footprint to ensure that the information available about you aligns with your brand (or "on brand"). Think carefully when deciding when, how, or whether to respond to criticism, even if the criticism is unfair. Sometimes, it is better just to let things go rather than give more life to comments that would have gone unnoticed if you had ignored them.

Position Yourself as a Thought-Leader (Regardless of Your Career Level)

Don't shy away from establishing yourself as a thought-leader in your field. If you have knowledge regarding a job role or industry or a better way to get work done, create and share meaningful information based on your thoughts, skills, and expertise. Compose articles, contribute to specialized publications, and discuss your findings and observations on various social media platforms. Your brand will be strengthened, and there will be opportunities for speaking engagements, media features, and collaborations as a result of your thought leadership content. It is also important to cultivate and maintain meaningful professional relationships with peers, mentors, industry leaders, and other influential figures in every forum— local, regional, national, and global, depending on your sphere of influence. Participate in activities designed specifically for networking, such as attending conferences and joining online forums. Then, share those experiences online.

Maintain Your Authenticity

Authenticity is when *who* you are, *what* you believe, and *how* you show yourself to the world are all in sync. Being authentic, honest, and transparent in your ideas, actions, and interactions is required. You can establish a powerful professional brand and build trusted relationships with colleagues, clients, and stakeholders if you embrace honesty and prioritize it in your life. Maintaining authenticity by adhering to your core beliefs and pursuing your interests meaningfully is essential to developing a powerful and long-lasting professional brand. Ensure that your brand is an authentic portrayal of who you are by maintaining consistency with

Valerie Capers Workman, Esq.

your core beliefs. But remember that being yourself on professional platforms does not necessarily mean you must be *all* of yourself— just the parts in line with your career goals.

Curate Your Social Media

Almost all of who you are in the world can be found online. Today, tomorrow, or someday, you will either be thrilled at how you protected your brand or regret the time you'll need to spend remediating the ancient "off-brand" content that conflicts with your current and future career goals. Whether or not you have made peace with this fact, your online presence indeed tells your story. Therefore, you are responsible for ensuring that your story's only author and final approver is you.

THE SCIENCE: GET SET

"There is no fate but what we make for ourselves"
~ John Connor

Reverse Engineer Your Career

"I think I'm really taking a good look at the big picture. That's the difference between being around for the final or watching the final from my sofa at home."

~ Pele

Reverse Engineering: The 7th Superpower Required to Advance Your Career

W hen I was a National Manager at KPMG, my goal was to become a Chief Compliance Officer for a publicly traded company. I researched the backgrounds of professionals currently in that role and mapped out the various jobs they took on their way to that position. I noted that numerous roads could lead to this C-level function, and each road was paved with different job titles along the way. Having this big picture allowed me to make a series of seemingly incongruent moves that ultimately led me to the C-Suite. By conducting regression analysis on my target job, I was always making progress in my career even if, at the time, this progress was only visible to me.

Valerie Capers Workman, Esq.

In the age of AI, linear work-your-way-up-the-career-ladder thinking is no longer the best way to navigate the complex and dynamic workforce paradigm. To achieve long-term success and fulfillment, you must embrace big-picture thinking and utilize the principles of regression analysis to get where you want to go. By deconstructing your ultimate career goal and strategically planning and staying open to considering multi-directional career moves, you can create a game plan that liberates you from sequential thinking and allows you to make forward, backward, and lateral moves with purpose and confidence, knowing that each action is progress.

I. Understanding Reverse Engineering in Career Planning:
As it relates to career planning, reverse engineering involves starting with the end goal and working backward to identify the steps needed to achieve it. Applying this process to your career planning will allow you to envision your ultimate career destination and identify the required skills, experiences, and opportunities.

1. **Defining the Ultimate Career Goal**: The first step in reverse engineering a career plan is to envision the ultimate career goal. This requires introspection, self-assessment, and understanding your personal passions and aspirations. Decide the type of role you want. Do you want to be a manager of teams, or do you want to be an individual contributor? Read job descriptions that interest you. When you find the one that excites you, set that role as your target job. Aim as high as you want to go. You can always make changes and adjustments, but having a goal in sight is important. Otherwise, you could find yourself drifting aimlessly. Setting a clear vision allows you to align your actions with your long-term objectives and confidently make short-term, career-progressive moves.

2. **Identifying Key Milestones**: With the ultimate career goal in mind, you can identify key milestones or achievements necessary to reach that destination. These milestones serve as checkpoints along your career journey, providing a sense of direction and progress. Where are you now? Are you still on the right path? Have you stayed in one place too long? Are you considering leaving too soon? Tracking your milestones will help you know when and if you need to make a move. (In *Pitfall Mitigation,* I will discuss how to avoid the unforced errors that can derail your progress.)

3. **Outlining Necessary Skills and Experiences**: Reverse engineering will enable you to identify the skills, experiences, and knowledge you need to acquire to reach each milestone. Refer to your target job frequently. Which skills have you already acquired? Which skills do you need? This process helps you prioritize development areas and provides a roadmap for your skill enhancement.

II. Embracing Big-Picture Thinking:
Big-picture thinking involves seeing the larger context and broader implications of decisions and actions. In contrast to sequential thinking, which focuses on short-term outcomes, big-picture thinking frees you to explore multi-directional career moves that align with your long-term goals without fearing that you are moving in the "wrong" direction.

1. **Breaking Free from Sequential Thinking**: Sequential thinking often traps you in a linear career trajectory, where you will feel compelled to follow a predefined path without considering alternative routes. Big-picture thinking encourages you to break free from this constrained mindset

and explore diverse opportunities that will contribute to your long-term growth. Sequential thinking will tell you that you've been in the same role for two years, and it's time for a promotion or a new job. Big-picture thinking will tell you that you've been in the same role for two years, and it's time to acquire more new skills. That promotion might add absolutely nothing to your résumé besides a "better" title. If you are looking to earn more money for the same role, you are looking for a salary increase, not necessarily a promotion. It is perfectly fine to seek a promotion if you believe you earned the next-level title. But it's vital to know the difference between a promotion and a salary increase.

2. **Valuing Forward, Backward, and Lateral Career Moves**: By adopting big-picture thinking, you will recognize the value of lateral, forward, and backward career moves. Forward moves align with traditional career progression, while backward and lateral moves provide valuable experiences and skill development. To accelerate your career progression, you must view all three types of moves as potential stepping stones rather than detours. Refer to your target job to see how the potential move will track to determine whether a possible move is right for you—regardless of how the move will look to your family and your peers

3. **Mastering the Power of Multi-Directional Progress**:

   a. *Forward* career moves often align with typical notions of career progression, such as promotions or advancements within the same field or organization. While these moves can be rewarding and showcase your success, they may not always offer the best way to progress in your career.

b. ***Backward*** career moves may seem counterintuitive, but they can provide essential knowledge and skills that have a transformative impact on long-term career success. For example, transitioning to a lower-level position in a different department, industry, or company can broaden your skill sets and ultimately propel you forward. Stepping backward from Vice President at Realogy to director-level Counsel at Focus Brands was one of the best career moves I ever made. At Focus Brands, I was quickly promoted to Vice President, Legal. Quantum Progression personified.

c. ***Lateral*** career moves, often considered to lack the cachet of a promotion, can open doors to exciting opportunities for growth. Transitioning to a role that may not align directly with your current skill set allows you to diversify your experience, build a reputation for flexibility and versatility, and explore your interests outside the constraints of a linear path.

Reverse Engineering: Key Takeaways

By visualizing your end goal, embracing a multi-dimensional career path, and staying agile in the face of change, you will confidently navigate the ever-evolving, AI-driven job market.

First Principles to Determine Your Core Strengths

"For we do not think that we know a thing until we are acquainted with its primary conditions or first principles and have carried our analysis as far as its simplest elements."

~Aristotle

First Principles Strength Assessment: The 8th Superpower Required to Advance Your Career

As AI tools become more prevalent, you will need to focus on your unique strengths to stand out and excel in your career. First principles thinking is the ideal method to help you identify your innate work-related strengths and weaknesses. Knowing your best "work self" at this most basic level will enable you to pursue job opportunities that align with your strengths. By pursuing roles that play to your core, you will increase your ability to perform with excellence and add value to any organization. Not knowing exactly what makes you excellent at work is like being a superhero who has not yet discovered the extent of their powers. How formidable did Wonder Woman become when she discovered her core strengths?

First Principles thinking involves breaking down complex problems into their most fundamental parts. Instead of following pre-set conventions or mimicking what others have done, you deconstruct the issue to its basic elements. By applying this methodology to identifying your professional strengths, you can identify what you excel at and understand why you excel in those areas. This deep understanding can help you optimize your performance, capitalize on your unique strengths, and position yourself more effectively in your professional journey.

It is important to understand that the strengths that make you excellent are not necessarily the ones you most enjoy. This is why it is so important to pursue enjoyment outside of work. Know what makes you excellent and focus on pursuing roles that rely on your particular strengths. Certainly, choose work settings that are satisfying at companies where you are treated with respect and dignity. But know it's okay to *like* your job; you don't have to *love* it. Focusing only on jobs that allow you to do what you love is a luxury few will enjoy in the age of AI. Job satisfaction should be the goal for your work life; joy is for your real life.

As I noted earlier, when my boss at KPMG asked me why I was so good at my job, I disappointed us both when I could not answer that question. From that moment on, I was determined to better understand my strengths and learn why I was naturally excellent in certain functions while other tasks were more difficult. In the years that followed, I developed a process for self-assessment that allowed me to distill my work strengths down to their most basic elements. When I joined Tesla, I learned about First Principles and the benefits of applying this theory to problem-solving. As I began to see this process in action, I realized I had been using the First

Principles methodology to determine my core strengths. It was invigorating to put a framework to the self-assessment process that I developed.

At Tesla, I also realized this was the process I had used for years to help me identify undiscovered talent hiding in plain sight. I know how to spot excellence even when excellence does not see itself. And because I believe excellence is title and function agnostic, I know how talent can be leveraged in roles that the professional may never have contemplated as a possible career option. This ability to identify core strengths in myself and others is the most fulfilling of all my professional superpowers. I know I have changed lives by helping people understand the fundamentals of their excellence in order to learn *why* they are excellent and *how* they can leverage that excellence to advance their careers. Below, please find my thesis on a novel way to identify the basic elements of your excellence:

Utilizing a First Principles Approach to Identify Professional Strengths

Introduction

In the realm of professional development, numerous tools and strategies aim to help individuals ascertain their strengths, weaknesses, and areas of growth. These range from standardized tests, such as the Myers-Briggs Type Indicator (MBTI), to feedback-based tools like 360-degree evaluations. Yet, amid this vast array of tools, the first principles approach stands out as a particularly potent methodology, especially when the objective is to identify one's unique professional strengths. Let's dive into why the first principles methodology holds a distinct edge over other self-assessment tools.

Valerie Capers Workman, Esq.

1. The Essence of First Principles Thinking:
At its core, first principles thinking is about breaking down a concept or problem into its most fundamental truths. It encourages individuals to strip away assumptions, conventions, and historical precedents, compelling them to approach problems from a fresh and uninfluenced perspective.

2. Genuine Self-Reflection versus Standardization:

    a. **Customized and Relevant Insights vs. General Descriptions:** Traditional tools like MBTI pigeonhole individuals into certain predefined categories. While these categories might offer general insights, they can only capture some people's unique nuances and individualities. In contrast, the first principles approach promotes a deep, personalized introspection that isn't limited by predefined classifications.

    b. **Avoiding Confirmation Bias:** Standardized tests often inadvertently lead to confirmation bias, where individuals might align their perceptions with the generalized feedback they receive. By pushing for raw analysis, the first principles method minimizes the risk of such biases.

3. Flexibility and Adaptability:
Unlike static assessment tools that offer a one-size-fits-all analysis, the first principles approach is dynamic. It allows individuals to adapt their self-assessment based on changing circumstances, new experiences, or evolving professional landscapes. This flexibility ensures that professionals continually align their strengths with real-world demands.

4. Depth Over Breadth:
While other tools provide a broad overview, first principles thinking encourages deep dives into specific areas of one's professional life. Focusing on the root causes and foundational truths of one's strengths and competencies allows for a more profound understanding and appreciation of one's unique capabilities.

5. Encouraging Proactive Thinking:
Traditional self-assessment tools tend to be reactive. They offer feedback based on past actions or predefined parameters. In contrast, the first principles approach is inherently proactive. By deconstructing and analyzing, professionals are not just understanding their current strengths but are also better equipped to forecast future areas of growth or potential strengths they can develop.

6. Independence from External Validation:
Many conventional tools rely heavily on external feedback. While external perspectives are valuable, they may sometimes be colored by biases or limited observations. The first principles methodology emphasizes self-driven analysis, ensuring the individual's understanding of their strengths is intrinsic and authentic.

7. Integration with Real-world Challenges:
By its very nature, first principles thinking is about addressing and solving real-world problems. When applied to professional self-assessment, it ensures that the identified strengths are directly relevant to the challenges and opportunities in the professional world rather than being abstract or theoretical.

Valerie Capers Workman, Esq.

Applying First Principles Thinking to Determine and Leverage Your Strengths: A Step-by-Step Guide

1. **Define the Problem or Objective**:
   a. Clearly state what you're trying to understand or achieve. In this context, the problem is, "What am I truly excellent at in my professional life?"

2. **Deconstruct the Problem**:
   a. List every role, responsibility, and task you engage with in your work. No task is too small or insignificant. This exhaustive list provides a comprehensive view of your typical responsibilities.

3. **Systematically analyze each task:**
   a. What are the core components of this task?
   b. What skills are essential to execute this task effectively?

4. **Identify Fundamental Truths**:
   a. For each task, ask yourself:
      i. Which of these tasks do I find easy or intuitive, and why?
      ii. When do I feel most engaged or in 'flow' while working?

5. **Finding Patterns and Core Strengths**:
   a. Upon deconstructing each task, patterns will emerge. Perhaps analytical tasks come easily to you, or maybe you find that interpersonal interactions are where you shine. These patterns highlight inherent strengths. However, it's essential not to mistake 'ease' for

'enjoyment.' You might not enjoy a task that comes to you naturally. If you do the task well —with ease and excellence—the skills you use to complete that task are strengths regardless of whether or not you enjoy the work.

6. **Reconstruct from the Ground Up**:
   a. Using the fundamental truths and insights you've gathered, build a new understanding of your strengths.

7. **Apply and Test**:
   a. Use this new understanding to focus on projects or tasks where your strengths shine.
   b. Seek feedback: Ask colleagues or supervisors if they've noticed these strengths or if there are other areas you excel in that you might have missed.
   c. Refine your understanding: As you receive feedback and encounter new challenges, refine your understanding of your strengths. Feedback from colleagues, mentors, and supervisors offers an external perspective on your strengths. They can provide insights into areas where your capabilities shine, often in ways you might not recognize yourself. By gathering feedback, you can cross-reference your self-assessments and refine your understanding of your strengths.

8. **Iterate**:
   a. The process of understanding one's strengths is ongoing. Periodically return to the first principles approach to re-evaluate your strengths, especially after completing significant new projects, role changes, or receiving new feedback.

9. **Leverage Your Strengths**:
   a. Identifying your strengths is just the first step; the next is leveraging them for career growth.
      i. Task Allocation: If you're in a team setting, volunteer for tasks or projects that align with your strengths.
      ii. Professional Development: Pursue courses or certifications that enhance your identified strengths.
      iii. Career Navigation: If you're considering a job change, look for roles that align more closely with your strengths.

10. **Continuous Learning**:
    a. Core strengths can be improved and refined over time. Engage in professional development, training, or courses that further enhance your identified strengths.

11. **Addressing and Improving Weaknesses**:
    a. While the primary focus is on strengths, it is equally essential to acknowledge areas of improvement. Knowing your strengths doesn't mean ignoring your weaknesses. Instead, you'll better understand which weaknesses are essential to address and which are less critical.

12. **Continuous Reflection**:
    a. Roles, responsibilities, and your professional aspirations will change over time, particularly if you are in Quantum Progression mode. Regularly applying the first principles approach ensures that your understanding of your strengths remains relevant and accurate. Set aside time—perhaps

quarterly or biannually—to reassess your strengths, gather new feedback, and adjust your professional path accordingly.

First Principles Self-Assessment: Key Takeaways

While there's no denying the value that traditional self-assessment tools bring to the table, my first principles approach offers a depth, authenticity, and adaptability that's hard to match. In the ever-evolving professional landscape, where uniqueness and genuine understanding of one's strengths are paramount, the first principles methodology will be one of your most valuable tools. By frequently re-applying this methodology, you will ensure that you are aware of your strengths and prepared to leverage them optimally in real-world scenarios.

First principle thinking empowers you to identify your work-related strengths, enabling you to make career choices that align with your natural gifts and talents. By choosing roles that play to your strengths, you can experience job satisfaction while you excel. As AI tools become more prevalent, the professionals who excel in a range of roles that they can flex into and out of, based on their clearly identified and perfectly leveraged strengths, will continue to be indispensable assets in the workforce.

Assemble Your Interdisciplinary Team

"I'm trying to set up opportunities for myself so that I don't only have one outlet to go through and rely upon as far as a support system financially, emotionally and mentally."

~Busta Rhymes

Configure Your Support System: The 9th Superpower Required to Advance Your Career

A strong interdisciplinary support team is a critical asset for professionals focused on career advancement. The goal is to assemble a group of experts who can provide guidance, advice, and consultation whenever you need it. Each team member should have a specific area of expertise to address key aspects of your professional growth and development. Your support system team members won't actually work together and likely will not know each other—although it's fine if they do. This "team" is your group of go-to experts who will help ensure that you function as close to the top of your game in all areas of your work life. This configuration is not just for the rich and famous and does not necessarily require an outlay of large amounts of money. Each of your team members does not even need to be human. Some team members might be websites or newsletters. Many experts in the disciplines you need provide free resources on social media platforms and podcasts. Whether you spend zero dollars or a small fortune, you need "Team You" for your career to progress in the age of AI.

Quantum Progression "Front Five" of an Interdisciplinary Team

1. The Tech Expert
2. The Career Coach
3. The Motivational Resource
4. The Therapist
5. The Medical Practitioner

1. The Tech Expert: Keeping You Current on Technological Advancements[26]

In the age of AI, *all* jobs will be tech jobs. Keeping up to date on technological advancements related to the workforce is crucial to your career success. *The Tech Expert* on your interdisciplinary team is the go-to resource for all things tech, including AI tools, social media trends, advancements in data science, cryptocurrency developments, etc. Whether you currently work in the tech space or not, you must be aware of everything happening in tech. You must avoid contributing to your obsolescence. Staying current with the tech industry will help ensure your continued relevance in the workforce.

---

[26] At the time of the publication of this book, a great resource to get you started is TheInformation.com ("We write deeply reported articles about the technology industry that you won't find elsewhere.")

2. The Career Coach: Navigating the Career Landscape

The *Career Coach* specializes in helping professionals pursue opportunities, create compelling résumés, and master the interview process. You can find an excellent coach by seeking recommendations and researching career coaching professionals with a proven track record of successfully assisting individuals in advancing their careers. Your career coach will help you set career goals, help you align your strengths and skills with opportunities, and guide you in preparing for interviews to present your best self. However, be very careful here. Either seek out free resources or thoroughly research the background of your potential career coach before you hire them. There are some incredibly inept individuals posing as career experts. Your career coach should be able to provide you with at least five satisfied clients who have backgrounds and career goals similar to yours. Also, regularly evaluate your career coach's effectiveness in helping you reach your career goals. If their guidance falls short, find a new coach. Most importantly, throughout your career, you will change coaches as you excel and outgrow a particular coach and as your career aspirations evolve.

3. The Motivational Resource: Nurturing Resilience and Positivity[27]

The *Motivational Resource* on your interdisciplinary team will provide guidance in matters of faith, positivity, and resilience. Depending on your spiritual beliefs, this team member can be a pastor, clergy member, religious congregation, or a motivational speaker known for providing positive and uplifting support and inspiration. The motivational resource will offer guidance in

---

[27] The New Testament of the Bible works best for me, particularly the parts that speak about minding my own business and not judging other people.

maintaining faith, finding positivity in adversity, and nurturing resilience during challenging times in your professional journey. Be sure the motivational resource's guidance aligns with your spiritual and emotional needs. Seek motivational support elsewhere if the connection is not equally uplifting and beneficial.

4. **The Therapist: Taking Care of Mental Health**

   The *Therapist* on your interdisciplinary team will help you address any mental health concerns and help you maintain a healthy work-life balance. Research licensed therapists or mental health professionals with experience supporting individuals in demanding professional roles. Select a professional who understands and can relate to your whole self. The therapist provides a safe space for you to discuss challenges, manage stress, and develop coping strategies for maintaining mental well-being. You will need to regularly assess the therapist's effectiveness in supporting your mental health. If there is a lack of progress or connection, seek alternative mental health support.

5. **The Medical Practitioner: Prioritizing Physical Health**

   The *Medical Practitioner* on your interdisciplinary team will help you prevent avoidable health issues from impacting your career. Look for a reputable medical provider who focuses on preventive care and comprehensive health management. Ensure you get regular health check-ups, take preventive care measures, and address any health concerns as soon as possible to help ensure your well-being.

The Interdisciplinary Support Team: Key Takeaways

Building an interdisciplinary team comprised of a Tech Expert, Career Coach, Motivational Resource, Therapist, and Medical Practitioner will provide the support you need to excel in your career. Yes, you *can* function without having this team in place, but why just function in the age of AI when you can be excellent?

Create Your AI Toolkit

"We are currently preparing students for jobs that don't yet exist...using technologies that haven't been invented ...in order to solve problems we don't even know are problems yet."

~ Richard Riley

Your AI Toolkit: The 10th Superpower Required to Advance Your Career

AI is transforming how businesses operate, making processes more efficient and unlocking new possibilities across industries. From automating repetitive tasks to analyzing vast amounts of data, AI is revolutionizing the workplace. Understanding the importance of staying current on the AI tools available to professionals in your current or desired field of work is critical. By understanding the role AI plays in various sectors and embracing the rapidly evolving AI landscape, you will stay competitive and continue to progress in your career. Instead of fearing AI as a threat to your job security, you should view AI as an opportunity to enhance your performance. As AI becomes more integrated into workplaces, the concept of

Valerie Capers Workman, Esq.

human-AI collaboration will continue to gain traction. You must learn to collaborate effectively with AI tools, understanding their strengths and limitations. AI is constantly evolving, and new tools and techniques emerge regularly. You will need to embrace a lifelong learning mindset to stay ahead of the curve and remain adaptable in the face of technological change.

I. AI Usage in Various Job Categories

To get you started, here are some ways AI allows users to increase their proficiency, efficiency, and performance. This list is by no means exhaustive, but will illustrate that AI is already everywhere in the workforce.

AI in Marketing and Sales: AI-powered tools analyze consumer behavior and preferences in marketing, enabling targeted advertising and personalized campaigns. Sales professionals benefit from AI-powered customer relationship management (CRM) systems that offer valuable insights and predictive analytics.

AI in Finance and Accounting: AI streamlines financial analysis, risk assessment, and fraud detection in the finance sector. For accountants, AI tools automate data entry and reconciliation tasks, reducing errors and freeing up time for higher-value activities.

AI in Human Resources: In HR, AI facilitates talent acquisition and workforce analytics. AI-powered chatbots offer employees instant support, enhancing the HR service experience.

AI in Healthcare: In the healthcare industry, AI aids in medical diagnosis, drug discovery, and patient care. AI-powered devices and algorithms improve treatment outcomes and patient safety.

Quantum Progression

AI in Education: AI-driven personalized learning platforms cater to students' needs, enhancing educational experiences and outcomes. AI-powered tools assist teachers in creating adaptive and engaging learning materials.

II. Warning About AI Bias and Ethical Considerations

As professionals embrace AI tools in their respective fields, it is essential to understand that AI will present challenges for users related to privacy and bias concerns.

Ethical considerations in data usage: AI systems rely heavily on data, and the responsible use of data is essential. Professionals must ensure data collection and processing adhere to privacy regulations and ethical guidelines. Transparency in data usage and obtaining informed consent from users are critical to the responsible utilization of AI applications.

Recognizing the propensity for bias in AI algorithms: AI algorithms are only as unbiased as the data on which they are trained. Biased data perpetuates societal prejudices and discrimination. Professionals who use AI must be vigilant in *detecting and correcting biases* in AI models to ensure fair and equitable outcomes. I had the opportunity to watch the AI visual artist Grihmlord Graphic Arts/Audio Visuals create the gorgeous cover "photo" for this book using AI. Prompt after prompt, it was astonishing to witness how challenging it was to get the results to produce images of business professionals of color. The most difficult images to generate? Black "professionals" and Black "business" men and women. A ridiculously simple prompt, but the results were profoundly disheartening. It took hours to get the representation I was looking for. I would not have believed this if I had not watched this process myself.

III. Obtaining Certification in a STEAM Discipline

The intersection of Science, Technology, Engineering, Arts, and Mathematics (STEAM) is transforming industries and shaping career opportunities. (For the purposes of this section, "Arts" means AI-enabled Arts.) The rise of AI is also reshaping job requirements and demanding new skills from professionals in every sector. It is imperative that you engage in continuous learning to remain current. This is important even if you have an advanced degree and especially true if you do not have a degree in a STEAM discipline. AI tools that were cutting-edge yesterday may be outdated today. Professionals who fail to keep current with AI tools risk falling behind their peers. Those equipped with AI knowledge can outperform peers and secure better career opportunities. A lack of AI knowledge can hinder career advancement in a world where AI is becoming a standard in many industries. Employers seek candidates who can leverage AI to drive business growth and success. To stay current with AI tools, you must invest in AI education and training. Online courses, workshops, and certifications are readily available to help you develop AI proficiency.

IV. Debunking Misconceptions About STEAM in Non-Technical Professions

Contrary to common misconceptions, STEAM is not confined to technical industries. Non-technical professions benefit from STEAM knowledge. As outlined above, marketing professionals leverage data analytics to drive targeted campaigns, while HR professionals can use AI-powered tools for talent acquisition and retention.

V. The Impact of STEAM Certification on Career Advancement

STEAM-certified professionals are better positioned to lead cross-functional teams and drive tech-focused innovation. The ability to apply cross-disciplinary knowledge fosters a culture of creativity and problem-solving within organizations. In a competitive job market, STEAM will give you a distinct advantage.

VI. Embracing STEAM for All Career Levels

Hourly and salaried workers can benefit significantly from STEAM certification, as it opens doors to higher-paying roles and greater job security. In industries like manufacturing and logistics, enhanced skills in automation and technology are becoming essential for remaining competitive in the job market. From gaining proficiency in AI-powered products and services to developing data-driven strategies to implementing AI-powered problem-solving solutions, STEAM proficiency can lead to new job opportunities, promotions, and increased leadership opportunities.

AI Toolkit: Key Takeaways

Staying current on all AI tools is no longer an option; it is imperative for professionals who seek to excel in their careers. From entry-level to executive roles, AI proficiency empowers professionals to deliver exceptional results, optimize work processes, and remain competitive in the evolving job market. By embracing AI and continuously upgrading your STEAM skills, you will be well-positioned to thrive in an AI-driven workforce.

Valerie Capers Workman, Esq.

A Recitation of AI Disciplines

Here is a recitation, though not a complete list, of AI disciplines for your consideration...

Machine Learning; Introduction to Machine Learning; Deep Learning Fundamentals. Supervised and Unsupervised Learning; Reinforcement Learning; Natural Language Processing (NLP); NLP Fundamentals; Sentiment Analysis; Named Entity Recognition (NER); Machine Translation; Computer Vision; Image Recognition; Classification Object Detection; Facial Recognition; Image Generation with GANs (Generative Adversarial Networks); Robotics and Autonomous Systems; Robotic Process Automation (RPA); Autonomous Vehicles; Drones. Human-Robot Interaction; Data Science and Analytics; Data Preprocessing; Data Cleaning; Data Visualization; Feature Engineering; Predictive Analytics; AI Ethics and Bias: Ethical Considerations in AI; Bias Detection and Mitigation in AI Systems; Fairness and Transparency in AI Algorithms; AI for Business and Industry; AI Strategy and Implementation; AI in Marketing and Customer Experience; AI in Supply Chain and Logistics; AI for Financial Services; AI in Healthcare; Medical Imaging and Diagnosis; Electronic Health Records (EHR) Analysis; Drug Discovery with AI AI-Driven Personalized Medicine; AI and IoT (Internet of Things); AI in Smart Cities; AI-Enabled IoT Devices; Predictive Maintenance with IoT and AI; AI and Big Data Handling and Processing; Big Data with AI;AI for Data Warehousing and Data Lakes; Real-time Analytics with AI; AI in Gaming and Entertainment; AI in Video Games and Virtual Reality; AI-Generated Art and Music; Personalized Content Recommendation Systems; AI and Cybersecurity; AI-Enhanced Threat Detection; AI for Anomaly Detection; AI in Cyberattack Prevention and

Quantum Progression

Response; AI Development and Programming; Python for AI and Data Science; TensorFlow and Keras for Deep Learning; PyTorch for Neural Networks; AI Research and Trends: Staying Up to Date with AI Research Papers and Conferences; AI Blogs and Podcasts; Participating in AI Hackathons and Competition.

Liberal Arts Education + Tech Skills = Winning

"Knowledge is Power"

~ Original Author Unknown

Liberal Arts + Tech Skills: The 11ᵗʰ Superpower Required to Advance Your Career

As a result of the COVID-induced worker shortage, the debate over the value of a Liberal Arts education versus the value of job-specific skills training, including tech skills, began to dominate all conversations about the future of workforce recruiting. Some argue that a skills-focused background is the key to career success, while others emphasize the enduring significance of a Liberal Arts education. I am certain that a combination of Liberal Arts education and skills-based training, particularly tech skills, will be the winning combination to enable your career advancement in the age of AI.

Valerie Capers Workman, Esq.

The Enduring Value of a Liberal Arts Education

Critical Thinking and Problem-Solving
A Liberal Arts education emphasizes critical thinking, analytical reasoning, and problem-solving skills. These skills are essential to any career. Professionals who can analyze complex issues, think critically, and propose innovative solutions are highly sought after, regardless of industry.

Effective Communication
Effective communication is a cornerstone of success in any field. Liberal arts programs emphasize written and verbal communication, helping graduates articulate ideas clearly and persuasively. In leadership roles, the ability to communicate effectively is paramount. Even with the availability of sophisticated language models, such as ChatGPT developed by OpenAI, a professional must have the ability to understand, evaluate, edit, and fact-check the text that is generated from these types of AI models.

Ethical and Social Awareness
Liberal Arts programs often explore ethical and societal issues, fostering a sense of social responsibility. Professionals with a strong ethical compass are more likely to make sound decisions and navigate complex ethical dilemmas, which is essential in leadership positions.

Tech Skills In the Age Of AI

The Importance of Tech Skills
There's no denying that tech skills are in high demand. In the age of AI, industries across the board rely on technology for efficiency and innovation. Proficiency in tech skills can open doors to a wide range of career opportunities. Tech skills provide a competitive

edge in the job market. They enable professionals to harness the power of automation, artificial intelligence, and data analytics to drive efficiencies, make data-driven decisions, and drive business growth. Tech-savvy professionals are often at the forefront of innovation. Many entry-level positions in industries outside of tech now require specific technical skills.

Tech Skills Alone Are Not Enough
While tech skills are valuable, they are not a standalone ticket to the C-Suite. Leadership roles demand a broader skill set, including strategic thinking, leadership, communication, and a deep understanding of the industry. Tech skills, while crucial, provide only a portion of the skill set needed for executive positions. Most importantly, as of the writing of this book, and conceivably for decades to come, tech tools are only as useful as the subject matter expertise of the functional leaders managing their deployment and utilization.

The Synergy of Liberal Arts and Tech Skills

The Holistic Professional
Professionals who combine a Liberal Arts education with tech skills become holistic leaders. They can approach complex challenges with critical thinking, communicate effectively with diverse teams, and adapt to technological advancements. This holistic approach is invaluable in leadership roles. The synergy of liberal arts and tech skills bridges the gap between technology and humanity. In leadership, the ability to understand both the technical and human aspects of an organization is essential. Leaders who can translate technical insights into strategic decisions while fostering a collaborative and ethical work environment are highly effective. Innovation often arises at the intersection of disciplines. Professionals

Valerie Capers Workman, Esq.

with diverse skills can identify unique solutions to complex problems. The combination of a liberal arts mindset and tech skills can fuel innovation and drive a company's growth. Professionals who blend a liberal arts background with tech skills are well-positioned for career advancement.

Hypothetical Success Stories

Morgan: The Tech-Savvy CEO:
Morgan started their career as a software developer, leveraging their tech skills to rise through the ranks. However, it was their ability to communicate the value of technology to non-technical stakeholders, honed through their liberal arts education, that helped propel Morgan to the CEO position. Morgan's holistic approach allowed Morgan to lead their company through a successful digital transformation.

Avery: The CTO with a Philosophy Degree:
Avery's degree in philosophy may seem unrelated to their role as a Chief Technology Officer. Still, their philosophical training taught Avery to think critically and question assumptions. This mindset helped Avery make innovative decisions in tech, leading to the development of groundbreaking products and, ultimately, their appointment as CTO.

Liberal Arts + Tech Skills: Key Takeaways
In the age of AI and rapid technological advancement, the synergy of a Liberal Arts education and tech skills is a recipe for career success and longevity. A Liberal Arts education equips professionals with critical thinking, communication, adaptability, and ethical awareness—qualities that are timeless and essential for leadership roles. Tech skills, on the other hand, provide a competitive edge and

access to a broad range of career opportunities.

Professionals who genuinely embrace this combination are poised to have fulfilling careers and make a lasting impact in their workplaces and industries. Limitless career advancement will be available to those who can use technology effectively while staying true to the core values and skills that make us human. Blending these strengths will exponentially increase your prospects for career advancement in the age of AI.

A Synopsis of Skills-Based AI Knowledge Paths

The options for the wide variety of four-year and two-year higher education degrees are well-covered in a plethora of websites and resource books. Lesser-known options to learn AI skills include apprenticeships, boot camps, and self-study courses, which are discussed below.

APPRENTICESHIPS

Apprenticeships provide real-world exposure to AI projects, allowing you to build practical skills under the guidance of experienced mentors. This expertise is offered in the form of "hands-on experience." These programs accelerate your comprehension and use of AI principles in the workplace and allow you to put your new skills to work immediately.

Earning While Learning: Apprenticeship programs provide participants with a salary while learning. This alleviates any financial hardship that may be present and offers an opportunity for individuals from various socioeconomic backgrounds to acquire valuable knowledge regarding AI.

Opportunities for Employment: Many apprenticeship programs include relationships with industry-leading organizations, which increases the likelihood of landing high-paying positions once the program is finished. This direct relationship to or with employers can be an advantage when applying for upper-level employment, particularly in AI-driven companies.

Equity Considerations: Apprenticeship programs tend to actively strive to attract individuals from backgrounds where they are underrepresented in artificial intelligence-related industries. These programs help to overcome the Equal Access gap in the technology industry by offering practical skills and a supportive learning environment. Participants typically receive hands-on experience and mentoring from industry professionals.

Availability: Apprenticeship options in AI are limited, making it difficult to find and get accepted into programs that are appropriate for your needs. Because of this scarcity, access may be restricted, particularly for professionals who live in places with a lower concentration of tech industries or businesses focused on AI. In addition, the selection process can be difficult and demanding, and not all applicants will necessarily be accepted.

BOOT CAMPS

Fast-Paced Upskilling: You can gain a thorough understanding of AI applications relatively quickly by participating in intensive training programs known as boot camps. The intensive learning experience benefits you if you are interested in rapidly upgrading your skills and enables you to immediately implement AI concepts into your current or prospective job.

Opportunities for Expanding One's Network: AI boot camps typically facilitate relationships with industry professionals and possible employers. This allows you to broaden your networks and investigate AI's potential applications in various fields.

High Intensity: Because of the high level of commitment and concentration required, boot camps are not appropriate for everyone, particularly if you currently have employment obligations. If you are already juggling many demanding professional and personal obligations, you may find it difficult to commit the time and energy to effectively finish a boot camp.

INDEPENDENT STUDY

Self-directed study allows you to learn at your own speed, enabling you to balance your education with your present obligations better. You will have more control over when and where you study, making this a practical choice if you already have a lot on your plate.

You can tailor your education to focus on particular AI subfields pertinent to your career objectives and areas of interest. With this individualized approach, you can obtain knowledge in the facets of AI most pertinent to your work and immediately implement this knowledge.

Availability: Self-directed learners have a large selection of resources to choose from due to the availability of many online courses, tutorials, and open-source materials. This accessibility democratizes artificial intelligence education and makes it possible for anyone from various backgrounds to obtain AI expertise. Because there is no formal direction or deadlines, self-directed study requires an extremely high level of self-discipline and motivation. You need the drive and determination to keep moving forward as you juggle your personal and professional obligations.

Equity Considerations: Individuals from a wide range of cultural and socioeconomic backgrounds are able to attend AI education at their own pace, thanks to the availability of a flexible and inexpensive option provided by it. Internet resources and open-source materials also democratize AI learning by lowering entrance barriers.

However, some companies may place more importance on formal degrees or certifications, possibly bypassing candidates with self-taught skills. If a company prioritizes on more traditional educational qualifications, it could make it difficult for self-directed learners to obtain or rise to executive-level positions in that organization.

THE EXECUTION: GO

"There is a difference between knowing the path and walking the path."

~ Morpheus

Passive Pursuit

"If You Build It…"

~ Ray Kinsella

Passive Pursuit: The 12th Superpower Required to Advance Your Career

I ronically, one of the most effective ways to find your next great job is to use a process I call Passive Pursuit. Rather than sporadically, randomly, or abruptly searching for job opportunities, you should be utilizing the power of AI tools to create an online presence designed to draw recruiters to you. Creating a captivating online presence using AI tools can be a game-changer, allowing you to showcase your skills, expertise, and unique value proposition in a way that positions you as a top candidate for jobs you won't even know about until a recruiter contacts you. Taking this passive approach–ironically requiring frequent activity–is an ideal way to land your next role. You will never have to restart your job search because you will perpetually be "open" to opportunities. Passive Pursuit means you will never have to scramble to update your online presence. If you stay ready, you never have to *get* ready. In addition, if you are in Passive

Valerie Capers Workman, Esq.

Pursuit mode and need to find a new job quickly, gearing up to switch to active job search mode will be a faster and less stressful process.

PASSIVE PURSUIT

I. Crafting an Alluring Digital Brand

1. **Define Your Personal Brand**: This involves identifying your unique strengths, skills, and values and aligning them with your career aspirations. As discussed in Managing Your Brand, a clear and compelling personal brand will serve as the foundation for creating an authentic online presence that resonates with recruiters.

2. **Optimize Your LinkedIn Profile:** LinkedIn is a crucial platform for seasoned professionals seeking to attract recruiters. Optimize your profile with relevant keywords, showcase your accomplishments, and use a professional headshot to create a strong first impression. Engage with industry content, connect with thought leaders, and participate in relevant discussions to increase your visibility and credibility.

3. **Curate Your Social Media Presence**: Maintain a professional and cohesive social media presence across platforms like X, Instagram, and Facebook. Share industry insights, relevant content, and thought leadership to position yourself as a game-changer or an authority in your field. Avoid posting controversial content or engaging in unprofessional behavior that could deter potential recruiters.

124

II. Leveraging AI Tools for Job Market Insights

1. **Utilize Job Search AI Engines:** AI-powered job search engines offer valuable insights into the job market, allowing professionals to track hiring trends, demand for specific skills, and emerging job opportunities. Use these tools to gain a competitive edge and tailor your online presence to position yourself as an ideal candidate for the jobs that align with your career game plan.

2. **Research Companies and Industries:** AI tools can provide comprehensive research on companies and industries of interest. Use this information to tailor your online presence to appeal to specific employers and demonstrate your understanding of their needs and challenges.

3. **Analyze Competitor Profiles:** AI tools can analyze competitor profiles and job applications to identify the ideal profiles for professionals currently working in the jobs you are open to considering. Utilize these insights to enhance your online presence and set yourself apart from the competition.

III. Creating Engaging Content

1. **Start a Blog:** Creating and maintaining a blog focused on industry-related topics can be a powerful way to showcase your expertise and thought leadership. Consistently publish informative and engaging content that addresses industry pain points and offers solutions to your target audience.

2. **Engage in Professional Webinars and Podcasts:** Participate in webinars, podcasts, and online events as a

guest speaker to reach a broader audience and establish yourself as an industry expert. Use AI tools to identify relevant opportunities and connect with organizers.

3. **Share Case Studies and Success Stories:** Highlight your professional achievements through case studies and success stories on your website and social media platforms. Demonstrating tangible results you have delivered in past roles will boost your credibility and attract recruiters.

4. **Follow Your Current Company's Guidelines for Social Media Usage.** Be sure to stay current with and adhere to your company's guidelines and policies for posting content. Never reveal or discuss any company information whatsoever, and keep your content broad and general but interesting.

IV. Building a Network of Influencers

1. **Connect with Thought Leaders**: Build connections with influential professionals in your industry. Engage with their content, share their insights, and contribute to their discussions. Building strong relationships with thought leaders can elevate your online presence and attract recruiters.

2. **Join Industry Associations and Forums:** Participate in industry-specific forums and associations to connect with like-minded professionals and expand your network. Engaging in relevant discussions and sharing your expertise will help you gain visibility and attract recruiters.

V. Optimizing Your Online Presence

1. **Monitor Online Activity**: Regularly monitor your online presence and take prompt action to address any negative feedback or incorrect information. Your online reputation plays a significant role in attracting recruiters, so ensure it reflects your true capabilities.

2. **Stay Up to Date with AI Tools**: AI tools are continually evolving, and staying current with the latest advancements is essential. Regularly research and update your knowledge to leverage the most effective AI tools to enhance your online presence and attract recruiters.

Passive Pursuit: Key Takeaways

By crafting an alluring digital brand, utilizing AI tools for market insights, creating engaging content, building a network of influencers, and optimizing your online presence, you can establish yourself as a top candidate without actively applying for specific jobs. Embrace the power of AI tools to maximize your visibility and bring the recruiters to you. You might not be interested in pursuing the opportunities that come your way, but use these inquiries to expand your network and make meaningful connections with recruiters and potential employers.

Interview Games

"May the odds be ever in your favor."

~ Effie Trinket

Game the Interview: The 13th Superpower Required to Advance Your Career

Game theory strategies can provide a competitive advantage in interview settings, enabling you to make strategic decisions that maximize your chances of success. Game theory principles can be applied to help you crush any interview. By understanding the underlying dynamics of the interview process, anticipating the interviewer's moves, making informed decisions on how to approach your responses to questions, and using AI tools to help you prepare, you can effectively navigate the interview process and stand out as the ideal candidate for the position.

Valerie Capers Workman, Esq.

I. Understanding Game Theory

1. **Defining game theory:** Game theory is a mathematical approach examining strategic interactions between decision-makers. It analyzes the potential outcomes of individuals' choices in competitive situations, aiming to identify optimal strategies.

2. **Applying game theory to interviews**: An interview can be seen as a strategic game, where you and the interviewer make decisions that influence the overall outcome. By recognizing the strategic nature of interviews, you can employ game theory to anticipate the interviewer's moves and tailor your responses accordingly.

II. The Pre-Interview Phase: Preparation is Key

1. **Gathering information:** Before the interview, you should research the company, its culture, and the specific role you are applying for. Understanding the company's goals, values, recent achievements, and recent headlines will enable you to align your responses to interview questions with the organization's needs.

2. **Anticipating interviewer's objectives:** Game theory involves considering the motivations and objectives of the other players. In the context of interviews, you should analyze the needs and preferences of the interviewer to tailor your responses accordingly. Research each interviewer to understand their role in the company and anticipate how the position you are applying for might help them be more effective in their roles.

3. **Preparing for the Interview**: AI-powered tools offer virtual interview simulations where you can practice answering common interview questions and receive feedback on your performance. This helps build confidence and will help you refine your responses.

4. **Tailored interview scenarios**: Some AI tools allow you to choose specific interview scenarios based on industry or role, simulating realistic interview experiences and preparing you for different types of questions and situations.

5. **Analysis of job descriptions:** AI-powered tools can analyze job descriptions to predict the questions likely to be asked during an interview. By understanding the job requirements, you can anticipate and prepare for relevant questions.

6. **Identifying behavioral-based questions:** AI tools can help identify common behavioral-based questions, which focus on past experiences and how you handled specific situations. Preparing thoughtful responses to these questions is crucial.

7. **Non-verbal analysis**: AI tools can help you practice and provide you with an analysis of your facial expressions, tone of voice, and body language to assess your level of engagement and confidence. Understanding and managing your non-verbal cues can significantly impact interview outcomes.

Valerie Capers Workman, Esq.

III. Maximizing Utility in the Interview

1. **Strategic decision-making:** Game theory encourages strategic decision-making by analyzing possible scenarios and their potential outcomes. You can apply this principle by structuring your responses to demonstrate how your qualifications and experiences can add value to the organization in various scenarios.

2. **Using the 'minimax' strategy:** The minimax strategy in game theory involves minimizing the maximum potential loss. In the context of interviews, you can prepare for challenging questions and unexpected scenarios to demonstrate how you would be the safest, lowest-risk, and best choice for the role.

IV. Creating a Positive First Impression

1. **Optimizing the initial interaction:** In game theory, signaling is crucial. In interviews, you should strategically signal your confidence, enthusiasm, and professionalism through your body language, tone of voice, and overall demeanor.

2. **Crafting a compelling elevator pitch:** An elevator pitch is a concise, persuasive introduction highlighting your skills and unique value proposition. You can use game theory to develop a powerful elevator pitch that captures the interviewer's attention and leaves a lasting impression. Open with your elevator pitch (during the "tell me about yourself" phase of the interview) and very briefly close with a variation on this pitch (during your closing "thank you").

V. Addressing Behavioral and Situational Questions

1. **Game theory in answering questions**: Applying game theory to answer behavioral and situational questions involves considering the interviewer's objectives and aligning your responses with the organization's values and needs. This strategic approach ensures that you stand out as a potential asset to the company.

2. **Using the 'tit-for-tat' strategy:** The 'tit-for-tat' strategy involves reciprocating the other player's actions. In the context of interviews, candidates can use this principle to adapt their responses based on the interviewer's cues, showing empathy and building rapport. Successfully executing this strategy requires that you use active listening skills to ensure that you are listening with the intent to understand what is being said and not merely waiting for your chance to respond.

Gaming The Interview: Key Takeaways

Mastering interviews with game theory strategies empowers you to approach the process strategically, making informed decisions that maximize your chances of success. By understanding the principles of game theory, anticipating the interviewer's objectives, and tailoring responses strategically, you can crush any interview and stand out as the top choice for the position.

Champions, Allies, and Influencers

"If you want to go fast, go alone. If you want to go far, go together."
~ *African Proverb*

Champions, Allies, and Influencers: The 14th Superpower Required to Advance Your Career

To have a successful career in the AI age, you must ensure you have a work-based support system. Work accomplishments alone are not sufficient. Being excellent at what you do is critical, and I will cover excellence in the next chapter. But excellence without a support team to hold you up is like a tree falling in the forest.

This support team should consist of three groups: champions, allies, and influencers. Each group plays a vital role in shaping your career while with the company and even after you leave. This support network will help you optimize your potential for success by helping you navigate the company dynamics, identify internal opportunities to enhance your skills, expose you to valuable growth experiences, and ensure you receive credit for your accomplishments.

Valerie Capers Workman, Esq.

CHAMPIONS

Champions are more than mere colleagues; they are influential figures within the organization who recognize your unique abilities and potential. These individuals stand out in their readiness to advocate for you, pushing your professional boundaries and paving the way for your growth. Champions are typically found in mentors, supervisors, or seasoned colleagues who see beyond your present accomplishments and envision what you could become.

Champions within an organization are pivotal to unlocking a higher level of professional success. They see what others might overlook, advocate for you with conviction, mentor with wisdom, and stand firm in their belief in you. They are the unsung heroes behind many professional success stories, acting as catalysts that transform potential into reality. I have benefited from having champions advocate for me throughout my career, and I have championed numerous professionals, as well. One of the most rewarding aspects of ascending to the ranks of leadership is having the ability to speak for excellent talent when they are not in the room.

1. **Recognition of Potential**

   Champions are often among the first to identify and acknowledge your potential. They see the spark, the uniqueness that sets you apart from others. This recognition is not always tied to current performance but is more about recognizing your latent talent that, when nurtured, could lead to extraordinary achievements.

2. **Mentorship and Guidance**

   Champions take on a role that goes beyond mere advocacy. They become mentors, guiding you through the professional landscape, offering wisdom derived from experience, and providing insights that help you navigate complex situations. Their mentorship is not about telling you what to do but helping you find your path, aligning your career with your potential.

3. **Creating Opportunities**
   One of the most vital roles a champion plays is opening doors that might otherwise remain closed. They utilize their organizational influence and connections to create opportunities tailored to your skills and interests. This might include access to special projects, promotions, or introductions to influential individuals.

4. **Constructive Feedback**
   Champions are not mere cheerleaders; they are invested in your growth. As such, they provide constructive criticism and feedback that prompts you to look inward, identify areas for improvement, and strive for excellence. They do so with empathy and understanding, ensuring that the input serves as a catalyst for growth rather than a deterrent.

5. **Emotional Support**
   The path to professional growth is fraught with challenges and setbacks. Champions recognize this and offer professional support and encouragement to you during these testing times, helping you to see failures as learning opportunities and help guide you through recovery and resurgence.

6. **Modeling Success**
   Champions often act as role models, embodying the values, ethics, and professional standards you aspire to attain. Their success becomes a tangible example of what you can achieve, motivating you to pursue your dreams with renewed excitement.

7. Investment in Your Growth

Perhaps what sets champions apart the most is their genuine investment in your professional growth. They dedicate time, effort, and sometimes even resources to help you evolve. Their advocacy is not limited to words; they take actionable steps to ensure you receive the necessary resources to thrive.

Identifying a Champion

Finding a champion in the workplace requires keen observation and understanding of the organizational dynamics. Look for those who appreciate your work and are in a position to make or influence decisions regarding your career. However, you should not ask someone to be your champion. Be excellent, and it is highly likely a champion will reach out to you.

ALLIES

Allies are the teammates and colleagues who share your vision, offer moral support, and collaborate to drive success. While champions may guide your path and influencers shape your environment, allies are the ones who walk beside you daily, contributing to a cohesive and thriving professional life.

It is important to know that allyship is always inclusive and welcoming. Allies are about building teams where all are welcome, not creating exclusive clubs.

1. Moral Support

Allies understand your strengths and weaknesses and provide encouragement when you need it most. They build confidence by affirming your capabilities and help you bounce back from failures, fostering resilience.

2. **Shared Vision: Aligning Goals**
   Allies often share a common vision and align with your professional goals. This shared perspective creates a strong bond and motivates the team to achieve a common objective. The alignment of vision ensures that everyone is on the same page, reducing conflicts and enhancing collaboration.

3. **Collaboration: Synergizing Efforts**
   One of the defining features of allies is their readiness to collaborate. They understand that success is a collective effort and are willing to put aside personal interests for the team's benefit. This spirit of collaboration leads to a synergy where the whole becomes greater than the sum of its parts.

4. **Trust and Loyalty**
   Trust and loyalty form the cornerstone of the relationship between allies. They trust each other's abilities and remain loyal even in challenging times. This trust is not blind but built on mutual respect, transparency, and understanding, creating a secure environment where ideas can be shared openly without fear.

5. **Diverse Perspectives**
   Allies often bring diverse perspectives to the table, enriching the decision-making process and leading to innovative solutions. They challenge each other's ideas constructively, promoting creative thinking and helping in crafting strategies that are robust and multifaceted.

6. **Conflict Resolution: Navigating Challenges Together**
   When conflicts arise within a team, allies play a crucial role in resolution. They mediate, offer insights, and work towards

a solution that serves the team's interest. Understanding each other's personalities and values helps navigate conflicts with empathy and wisdom.

7. **Celebrating Success**
Allies not only work together during challenging times but also celebrate successes together. They amplify the joy of achievements, recognizing and appreciating each other's contributions. This celebration fosters a positive work environment and strengthens the bond among team members.

8. **A Culture of Empathy**
Allies humanize the workplace by creating a culture of empathy. They are sensitive to each other's needs, supportive in personal and professional matters, and act as a support system that transcends mere professional collaboration.

Allies in the workplace are not merely colleagues; they are the companions who make the journey enjoyable and rewarding. They transform the professional environment into a space where creativity thrives, collaboration is seamless, and success is a shared endeavor. Investing in building strong alliances within the workplace is not merely a strategic move; it's a human-centric approach that recognizes the essence of teamwork and shared values. In a world often driven by competition, allies remind us that collaboration, empathy, and shared vision are equally potent forces that drive success.

INFLUENCERS

In any organization, influencers play a vital role in shaping decisions and affecting the trajectory of initiatives and projects. Their broad network and subtle authority make them essential to recognize and engage with, particularly for professionals looking to advance their careers. Here's why influencers are crucial to career success within a company and how you can gain their support.

1. **Building Relationships**
   Connecting with influencers helps build a robust professional network, leading to more opportunities for collaboration and growth.

2. **Mentorship and Guidance**
   Influencers often have a wealth of experience and knowledge that can be tapped into for professional development.

3. **Enhancing Visibility**
   Being aligned with influencers can elevate your profile within the organization, making you more visible to senior leadership.

4. **Strategic Alignment**
   Influencers can help align your projects with the company's strategic goals, ensuring your efforts are recognized and rewarded.

How to Gain the Support of Influencers

1. **Identify the Influencers**
   Recognize who the influencers are within your organization. They may not hold high titles but are often key decision-makers or opinion leaders.

Valerie Capers Workman, Esq.

2. **Build Genuine Relationships**
   Don't approach influencers solely for personal gain. Focus on building a genuine relationship based on shared interests and values.

3. **Provide Value**
   Understand what motivates the influencers and how you can provide value to them. This could be through unique insights, project collaboration, or supporting their initiatives.

4. **Seek Guidance**
   Approach influencers for advice and mentorship. Show appreciation for their expertise and be open to their feedback.

5. **Communicate Clearly**
   Be clear about your goals and how the influencer can support you. Transparent communication fosters trust and collaboration.

6. **Show Respect for Their Time and Position**
   Influencers are often busy individuals. Being respectful of their time and appreciating their position within the company goes a long way in building a positive relationship.

7. **Reciprocate Support**
   Influencers are more likely to support you if they see you're also willing to lend your support to their projects and initiatives.

Champions, Allies and Influencers: Key Takeaways

Embracing these support systems is not just an intellectual exercise but a practical guide to navigating the multi-dimensional labyrinth of AI in the workforce. With a team of champions, allies, and influencers, career advancement is eminently achievable.

Continuous Excellence

"Celebrate what you've accomplished but raise the bar a little higher each time you succeed."

~ Mia Hamm

Continuous Excellence: The 15th Superpower Required to Advance Your Career

In the age of AI, maintaining continuous excellence at work is not only important but essential for everyone seeking to thrive in their careers. By consistently striving for improvement and embracing a mindset of constant learning, you will unlock your full potential and fuel the continued advancement of your career. Continuous excellence will set you apart from the competition—human and AI—ensuring your relevance, value, and long-term success in the face of technological advancements.

Valerie Capers Workman, Esq.

The Benefits of Excellence

Career Advancement

Maintaining continuous excellence enables you to seize new opportunities and confidently navigate career transitions. In an AI-driven world, job roles may evolve, new industries may emerge, and established sectors may undergo significant transformations. By fostering and owning your excellence, you will be better equipped to pivot into emerging fields and adapt to the changing demands of the job market. This adaptability will safeguard your career advancement.

Internal Mobility

Consistently delivering excellence is also an effective strategy for gaining recognition and advancing within an organization. Employers seek to retain and promote individuals who consistently exceed expectations and contribute significantly to the organization's success. When you consistently exemplify excellence, you are more likely to be entrusted with leadership responsibilities, offered high-impact projects, and considered for promotions.

Job Satisfaction

Continuous excellence fosters a sense of purpose and personal fulfillment in your career. Achieving excellence in your field and consistently delivering exceptional results instills a profound sense of pride and satisfaction in the work accomplished. This intrinsic motivation becomes a driving force that fuels your commitment to excellence and inspires you to go above and beyond.

Job Stability

In the AI era, excellence is also intrinsically linked to employability and job stability. As automation disrupts traditional job roles, those demonstrating continuous excellence stand a better chance of weathering industry shifts and economic uncertainties. While no job is guaranteed, employers are more likely to retain top performers, even during challenging times, recognizing the tremendous value they bring to the organization.

EXCELLENCE IN ACTION

1. **Cultivating a Passion for Learning**
   a. Knowledge expansion: Engage in continuous learning by reading industry publications, attending webinars or workshops, and participating in professional development programs.
   b. Skill diversification: Continually expand your skill set by acquiring new competencies, cross-functional opportunities, and challenging assignments that broaden your expertise.

2. **Embracing Failure as a Stepping-Stone to Success**
   a. See failure as a valuable learning experience, allowing you to identify areas for improvement and adjust your approach. Cultivate resilience and the ability to bounce back from setbacks.
   b. Embrace an iterative approach, continuously refining your skills and strategies based on feedback and outcomes. View failure as an opportunity to iterate and enhance your performance.

3. **Set S.M.A.R.T. Goals**
   a. Specific: Set specific goals that define what you want to achieve at work. Clearly outline the desired outcome and the steps required to reach it.

b. Measurable: Establish measurable criteria to track your progress and assess success. Define key performance indicators (KPIs) that align with your goals and allow for objective evaluation.

c. Attainable: Set goals that are challenging yet attainable with effort and commitment. Avoid setting too easy or unattainable goals, as they may lead to complacency or demotivation.

d. Relevant: Ensure your goals align with your career aspirations and organizational objectives. This alignment ensures that your efforts contribute to broader success.

e. Time-bound: Set deadlines and establish a timeline for achieving your goals. This time constraint fosters a sense of urgency and helps you stay focused and accountable.

4. Set Stretch Goals

a. In addition to achievable goals, set stretch goals that push you beyond your comfort zone and encourage continuous improvement. These ambitious goals drive innovation and professional growth.

b. Embrace the KAIZEN[28] philosophy of continuous improvement, consistently making small, incremental changes to enhance your performance and achieve excellence over time.

---

[28]"Over 30 years ago, Masaaki Imai sat down to pen the groundbreaking book *Kaizen: The Key to Japan's Competitive Success* (McGraw Hill). Through this book, the term KAIZEN™ was introduced in the western world." (Kaizen.com)

5. Cultivating High-Performance Habits

1. Effective Time Management:
a. Identify and prioritize tasks based on their importance and urgency. Focus on high-value activities that align with your goals and contribute to overall success.
b. Explore productivity techniques such as the Pomodoro Technique[29], time blocking, and task batching to optimize your time and maximize efficiency.

2. Proactive Self-Development:
a. Regularly reflect on your performance, identify areas of improvement, and develop action plans to enhance your skills and competencies.
b. Actively seek feedback from colleagues, supervisors, and mentors. Embrace constructive criticism and use it as a catalyst for growth and improvement.

3. Leveraging Technology for Excellence:
a. Utilize AI-powered tools to automate repetitive or time-consuming tasks, allowing you to focus on higher-value activities. (Use only company-approved resources.)
b. Implement AI-enhanced workflow management tools that streamline processes and improve efficiency, reducing the risk of errors and delays.

---

[29] The Pomodoro Technique: An Effective Time Management Tool, by Amrita Mandal, PhD (science.nichd.nih.gov/May 2020)

4. Data-Driven Decision-Making:

a. Data analytics: Leverage data analytics tools to analyze and interpret large datasets, gaining valuable insights for informed decision-making and process optimization.

b. Predictive analytics: Utilize predictive analytics tools to anticipate future trends, identify potential challenges or opportunities, and make proactive decisions.

5. Purpose Driven Work

a. Identify the intrinsic and extrinsic motivators that drive your job satisfaction. Align your responsibilities with your values and find purpose in what you do.

b. Ensure that your company's goals align with your values and contribute to your overall fulfillment and satisfaction.

6. Work-Life Integration:

a. Establishing boundaries: Set clear boundaries between work and personal life, allowing for dedicated time for relaxation, hobbies, and self-care.

b. Well-being practices: Prioritize self-care activities such as exercise, mindfulness, and regular breaks to maintain physical and mental well-being.

Continuous Excellence: Key Takeaways

Achieving continuous excellence at work requires a combination of a growth mindset, ambitious goal-setting, high-performance habits, and the strategic use of technology. By embracing a passion for learning, setting clear goals, cultivating high-performance habits, and leveraging technology effectively, you can unlock your full potential and strive for continuous improvement. Sustaining motivation and maintaining work-life balance are also crucial for long-term success and well-being. By adopting these strategies and integrating them into your professional journey, you can achieve continuous excellence, stand out in your career, make a lasting impact in your chosen field, and enable your career advancement.

Superhuman

"Good afternoon, gentlemen. I am a HAL 9000 computer."
~ HAL 9000

Superhuman: The 16th Superpower Required to Advance Your Career

While AI undoubtedly possesses remarkable capabilities, the idea that it will entirely replace human contributions is a misconception. From emotional intelligence to creativity, adaptability, and ethical decision-making, humans possess intrinsic attributes that ensure our superiority over AI in the workplace.

Advantage: Humans

I. Emotional Intelligence
Emotional intelligence (EQ) is a quintessential human trait that AI cannot replicate. Understanding and managing emotions are essential for effective communication, conflict resolution, and

building strong interpersonal connections. In business and industry, emotional intelligence is crucial for leadership, customer relations, and team dynamics. AI algorithms cannot replace the ability to empathize, motivate, and inspire colleagues.

II. The Creative Spark

Creativity is an unparalleled human attribute that fuels innovation and problem-solving. While AI can process vast amounts of data and recognize patterns, true creativity requires the capacity to think outside the box and connect disparate ideas. Human imagination drives the development of groundbreaking products, services, and strategies that push businesses forward.

III. Adaptability and Flexibility

In a rapidly changing world, adaptability is a crucial differentiator for humans in business and industry. Humans possess the cognitive ability to learn new skills, embrace change, and pivot when necessary. AI algorithms may be refined to perform specific tasks, but humans have the versatility to tackle a wide range of challenges and roles.

IV. Ethical Decision-Making

Ethical decision-making requires a moral compass and a deep understanding of complex social and cultural contexts. While AI can analyze data to optimize outcomes, it lacks the inherent sense of right and wrong that guides human judgment. Humans can consider multiple perspectives, weigh ethical implications, and make decisions that align with shared values and societal welfare.

V. Comfort with Ambiguity

Dealing with complex and ambiguous situations is an area where human intuition and experience excel. Humans possess the capacity to assess multifaceted challenges, identify alternative solutions, and integrate various factors to arrive at comprehensive decisions. Human expertise is crucial for navigating intricate scenarios.

VI. Emotional Connections with Customers

In customer-centric industries, the human touch is indispensable. Building emotional connections and establishing customer trust creates loyal relationships that AI cannot replicate. Human customer service representatives can provide empathy and personalized solutions, enhancing customer satisfaction and loyalty.

VII. Resilience and Grit

The ability to persevere through challenges and setbacks is an inherent human quality that drives success in business and industry. Humans can withstand adversity, learn from failures, and emerge stronger and more determined. This resilience fosters innovation and progress.

VIII. Intuition and Gut Instinct

Human intuition, also known as gut instinct, is a powerful tool in decision-making. It involves drawing on tacit knowledge, experience, and a deep understanding of the situation. While AI can process vast amounts of data, it lacks the intuition that allows humans to make quick, nuanced judgments.

IX. Leadership and Vision

Leadership is a distinctively human trait that inspires and motivates teams. Influential leaders possess vision, emotional intelligence, and the ability to communicate a compelling narrative that mobilizes others toward a common goal. AI may be useful in *supporting* leaders, but it cannot replace the vision and inspiration that great human leaders bring to their organizations.

X. The Human-Centered Approach

At its core, business and industry are about serving people. Humans

Valerie Capers Workman, Esq.

are uniquely qualified to understand the nuanced needs and desires of customers, employees, and stakeholders. The human-centered approach to business fosters meaningful connections and ensures business strategies align with human values and aspirations.

Superhuman: Key Takeaways Part One – The Superiority of Humanity

While AI has undoubtedly made significant strides in transforming various aspects of business and industry, the notion of complete automation and replacing humans is science fiction. The inherent qualities that define humanity—including emotional intelligence, creativity, adaptability, ethical decision-making, and intuition—will always set humans apart and guarantee our superiority in the workplace. Embracing the synergies between humans and AI can lead to transformative innovations and an enhanced human experience in the future of work. As we navigate the AI-driven landscape, recognizing and harnessing the unique strengths of humans and AI is essential to driving sustainable success and creating a more enriching and inclusive future for everyone.

Empathy: Your Competitive Edge

Leadership traits have adapted to changing societal, economic, and technological landscapes over time. In the industrial age, leadership often emphasized authoritarian control and efficiency. As workplaces became more diverse and knowledge-based, leadership traits shifted towards collaboration, communication, and adaptability. With the advent of the AI-driven Fourth Industrial Revolution[30], I believe

---

[30] The Fourth Industrial Revolution refers to a period of rapid and transformative technological changes that are reshaping various aspects of society, including the economy, industries, and the way we live and work. This revolution builds upon the digital revolution of the late 20th century but is characterized by the convergence of multiple technologies, including artificial intelligence, robotics,

leadership is facing another transformative shift. As automation and AI systems take center stage and the skilled worker shortage begins to impact productivity[31], companies will ramp up their efforts to attract and retain top human talent.

At the same time, the workforce of the future will recognize that empathetic leaders are best equipped to support them through this transformative era, providing emotional support, fostering a culture of inclusivity, and preserving the essence of humanity in the workplace. Consequently, the workforce will increasingly favor companies that prioritize empathy as a leadership trait, recognizing that this type of leadership will enhance their overall well-being and prioritize their professional growth.

Empathy, the ability to understand and share the feelings of others, is a quality that sets humans apart from machines. In the age of AI, where automation and algorithms play an increasingly significant role, the human touch will become a rare and invaluable commodity. Empathetic leaders possess the capacity to connect with employees on a deeper level, understanding their needs, concerns, and aspirations. This connection leads to improved communication, trust, and collaboration within the organization.

Empathetic leadership is not a mere concept—it's a set of behaviors and practices that demonstrate genuine care for employees' well-being. It will be a valued skill in leaders at every level, from the front line to the C-Suite. In the age of AI, empathy will become synonymous with excellence and will be the core competency needed to ensure your continuous career advancement.

---

the Internet of Things (IoT), 3D printing, biotechnology, and more. "The Fourth Industrial Revolution" Klaus Schwab, the founder and executive chairman of the World Economic Forum (WEF). (Penguin Group, 2017)

[31] "By 2030, demand for skilled workers will outstrip supply, resulting in a global talent shortage of more than 85.2 million people" (The Future of Work, Talent Crunch, Korn Ferry, Spring 2018)

Valerie Capers Workman, Esq.

Empathetic Leadership in Practice

1. **Active Listening:** Empathetic leaders engage in active listening to understand employees' concerns, ideas, and feedback. They prioritize comprehension before being understood.

2. **Open Communication:** They cultivate environments that encourage open and honest communication. This transparency builds trust and establishes a sense of psychological safety.

3. **Support and Development:** Empathetic leaders invest in their team members' professional development and growth, aligning individual goals with organizational objectives.

4. **Recognition and Appreciation:** They acknowledge and appreciate employees' contributions, celebrating their achievements and milestones.

5. **Conflict Resolution:** Empathetic leaders handle conflicts with sensitivity and fairness, seeking resolutions that uphold the dignity of all parties involved.

6. **Emotional Intelligence:** They exhibit awareness of their own emotions and those of others, using emotional intelligence to navigate complex interpersonal dynamics.

7. **Crisis Management:** During challenging times, empathetic leaders guide their teams with empathy and compassion, helping employees cope with change and uncertainty.

Empathetic leaders not only attract and retain top talent but ensure that employees are motivated and engaged, reducing the risk of turnover. Moreover, empathetic leadership is essential for managing challenging workforce decisions. In a talent-short environment where difficult choices will undoubtedly arise, empathetic leaders will make critical decisions with compassion and dignity, minimizing the impact on employees.

Superhuman: Key Takeaways Part Two – The Competitive Advantage of Empathy

As AI reshapes industries and the nature of work, leadership is evolving to emphasize qualities that are distinctly human. Empathetic leadership, marked by understanding, compassion, and emotional intelligence, is becoming increasingly vital. Companies that prioritize and cultivate empathetic leadership are better positioned to attract, retain, and empower top talent, particularly in the face of the looming talent shortage. Furthermore, they create work environments where employees experience grace and dignity during challenging times. In the age of AI, empathy is not just a desirable human trait; it's a competitive advantage that will define your professional excellence and help to propel your career advancement.

Lift as You Climb

Lifting As You Climb: The 17th Superpower Required to Advance Your Career

The "Lift As You Climb" concept, birthed by Mary Church Terrelle, holds immense significance in the fast-paced and competitive workforce landscape. It embodies extending a helping hand to others as you progress in your career. To maintain human superiority, it will be incumbent upon each of us to foster a supportive, mentoring, and empowering culture. As the impact of AI continues to shape workplaces in every industry, the human quality of caring for others stands as a fundamental aspect that sets us apart from AI. From nurturing future leaders to creating a collaborative and inclusive work environment, embracing the "Lift as You Climb" philosophy benefits individuals and leads to a stronger, more resilient workforce. If you have not already done so, seriously consider adopting this philosophy as one of your professional north stars.

Valerie Capers Workman, Esq.

I. The Power of Mentorship

Mentorship is a transformative tool that enables professionals to share their experiences, insights, and knowledge with aspiring individuals. As you progress in your career, serving as a mentor can guide others through the challenges and opportunities you have faced. By nurturing talent and providing guidance, you elevate others' careers and foster a sense of purpose and fulfillment in your professional journey.

II. Paying it Forward

The "Lift as You Climb" principle is rooted in paying forward the support and opportunities you have received in your career. Just as someone may have opened doors for you, offering support to others creates a ripple effect of positive change and empowerment. By lifting others, you become an agent of transformation in their lives and enable them to reach new heights in their careers.

III. Creating a Supportive Work Culture

Incorporating the "Lift as You Climb" philosophy into workplace culture nurtures collaboration, empathy, and camaraderie. As a leader or team member, fostering a culture of support and mutual respect enhances job satisfaction, boosts productivity, and encourages creative problem-solving. In contrast to AI, which lacks emotional intelligence, human-to-human support fosters a sense of belonging and motivation that transcends robotic efficiency.

IV. Equal Access and Inclusion

Supporting others as you progress in your career plays a vital role in promoting Equal Access and inclusion in the workplace. By lifting diverse voices and empowering individuals from various backgrounds, you contribute to a more inclusive work environment.

162

Embracing Equal Access leads to innovative solutions, improved decision-making, and a broader perspective on problem-solving, aspects that AI algorithms will struggle to replicate.

V. Nurturing Future Leaders

Identifying and nurturing potential leaders is an essential aspect of "Lift as You Climb." As you grow in your career, taking the time to mentor and develop future leaders ensures the continuity of excellence and a legacy of impact. Cultivating leadership qualities in others will shape your organization's future, helping foster a continuum of excellence based on your accomplishments and impact.

VI. Promoting Social Responsibility

The "Lift as You Climb" philosophy extends beyond individual career trajectories. As professionals, we have a responsibility to contribute positively to society. By uplifting others, we create a culture of social responsibility that goes beyond individual achievements. In contrast to AI, which operates without consciousness, human empathy drives social responsibility, leading to meaningful contributions to the greater good. Particularly in mission-driven companies, this emphasis on social responsibility will help you to exemplify your company's ability to walk the talk, thereby establishing you as a model leader—regardless of your level in the company.

VII. Cultivating Resilience

Lifting others and supporting their growth enhances your resilience, as well. Engaging in mentorship and collaboration fosters emotional intelligence, communication skills, and adaptability—vital qualities in navigating the evolving professional landscape. As AI continues to transform industries, resilience becomes a key differentiator for human professionals.

VIII. The Legacy of Impact

"Lift As You Climb" allows you to leave a lasting legacy of impact in the lives and careers of others. By contributing to the success of others, you build a network of support and create a web of positive influences that extends beyond your immediate sphere of influence. Accordingly, your sphere of influence will expand as those you support move on and advance in their careers, either internally or in new roles at other companies.

IX. Personal Fulfillment and Purpose

The act of supporting others brings profound personal fulfillment and purpose. Knowing you have played a role in someone else's success and growth enriches your professional journey. This sense of purpose transcends monetary gains and elevates your career satisfaction to a higher level.

Lift as You Climb: Key Takeaways

"Lift as You Climb" encompasses the profound impact of supporting others as you progress in your career. Embracing this philosophy creates a culture of mentorship, collaboration, and inclusion that sets us apart from the limitations of AI. By nurturing future leaders, promoting Equal Access and inclusion, and fostering a supportive work environment, professionals contribute positively to their own growth and the advancement of society. Lifting others not only enriches individual careers but also amplifies humanity's collective impact, demonstrating that the human touch remains indispensable in an AI-driven world.

Pitfall Mitigation

Staying too long, leaving too soon, or doing too much are three likely mistakes you could make when wrestling with the fear of AI's impact on your career. Stay too long, and you could miss growth opportunities. Leave too soon, and you could damage valuable relationships or squander the brand capital you have gained. Do too much, and you could overburden yourself with responsibilities that technology could handle more efficiently.

Moreover, you can sidetrack your career by making unforced errors. Two of the most common self-inflicted career obstructions I have seen are work issues caused by Imposter Syndrome and Unconscious Bias.

This section contains a brief overview of these potential pitfalls and how to avoid them.

Staying Too Long

The comfort and familiarity of staying at one job for an extended period can be enticing, but it can also lead to missing out on valuable career advancement opportunities. Understanding the potential career stagnation that can result from staying too long at one job is crucial.

While prolonged tenure at one company can showcase loyalty and commitment, it may also limit opportunities for skill development and growth. You could find yourself stuck in a role where you have mastered their tasks but have limited exposure to new challenges and learning experiences. The longer you stay in one job, the more likely you are to become specialized in your current role. While specialization can be beneficial, it can also hinder career advancement if you develop skills that are too narrow or specific to your current position. This can make it challenging to transition to new roles that require a broader skill set.

Staying too long in the same job can lead to career plateauing. Without new challenges and opportunities to stretch your abilities, you may discover your career growth stagnates and your motivation wanes. Career plateauing can lead to decreased job satisfaction and a lack of enthusiasm for work. You may feel unfulfilled and disengaged, which can negatively impact your performance and productivity.

Fear of Change

Change can be intimidating, and the fear of stepping into the unknown can cause you to become complacent in your current role. The security of a familiar environment, colleagues, and job responsibilities can make it difficult to contemplate leaving and embracing change. The fear of change is a natural human response. It stems from the uncertainty and unpredictability associated with new situations. Many professionals are content with their current roles because they have grown accustomed to the routine, know what is expected of them, and have established a level of comfort with their coworkers and workplace dynamics.

This attachment to the status quo can be likened to a cozy cocoon, shielding individuals from the uncertainties that change may bring. It's akin to staying within a well-defined comfort zone, where the challenges and surprises of new opportunities are held at bay. Yet, it's crucial to recognize that personal and professional growth resides just beyond the boundaries of this comfort zone.

The security of familiarity is compelling, and it is human nature to seek stability. However, it's also important to remember that embracing change can lead to personal and career development that might not be achievable within the confines of the familiar. While the fear of change is natural, it's not insurmountable.

One way to approach this fear is by acknowledging that stepping into the unknown can be an opportunity for growth, learning, and new experiences. Change can stimulate creativity, invigorate your professional journey, and open doors to unexplored possibilities. In these moments of transition and transformation, individuals often discover hidden potential and strengths they never knew they possessed.

While feeling apprehensive about change is perfectly normal, it's equally important to balance that fear with a sense of curiosity and a willingness to take calculated risks. After all, the ability to adapt, evolve, and embrace change often leads to the most significant personal and professional achievements. So, while the allure of the familiar is strong, the allure of the unknown can be even more rewarding if we dare to step outside our comfort zones.

The Comfort Trap

Staying at one job for an extended period can create a comfort trap. You may settle into a routine, become comfortable with the status quo, and inadvertently resist exploring new career opportunities beyond your comfort zone. The comfort trap can lead to a sense of complacency, where you become content with the familiar and resist taking risks. You may avoid seeking new challenges, which can limit your growth and potential in your career.

Staying too long in the same job can also lead to career plateauing. Without new challenges and opportunities to stretch your abilities, you may discover your career growth stagnates and your motivation wanes. Career plateauing can lead to decreased job satisfaction and a lack of enthusiasm for work. You may feel unfulfilled and disengaged, which can negatively impact your performance and productivity.

CONSIDER AN INTERNAL MOVE

Recognizing the need for change is the first step toward embracing new opportunities that will foster your career advancement. Self-reflection is essential for career development. If your current job no longer aligns with your objectives, it may be time to consider a change. Before you leave your current company, you should first consider making an internal move.

Assess Internal Opportunities

Exploring internal opportunities within the same company can be an effective way to break free from the comfort trap while mitigating potential career inertia. Internal job postings are an excellent resource for professionals looking to explore new opportunities within their current company. Review these postings regularly to ensure you do not miss any opportunities that might appeal to you.

Leverage Company Knowledge

An internal move will allow you to leverage your existing knowledge of the company's culture, values, and processes. This knowledge will give you a unique advantage in your new role, allowing for a smoother transition and faster adaptation.

Build on Relationships

If you have spent significant time in one company, you have likely built strong relationships with a finite set of colleagues. While these relationships are undoubtedly valuable, moving into a new role could expose you to an entirely new group of colleagues, thereby expanding your sphere of influence internally.

Expand Your Skill Sets

An internal move can present opportunities to acquire new skills and knowledge while building on existing expertise. By taking on new responsibilities or working in different departments, you can broaden your skillset and make yourself more marketable for future opportunities – internal and external.

CONSIDER AN EXTERNAL MOVE

You should evaluate your long-term career aspirations and consider whether your current role aligns with your goals. If your aspirations have outgrown your current job, it may be time to explore new opportunities outside of your current company. Evaluating your career aspirations requires introspection and long-term planning. Consider where you see yourself in the next three, five, ten, or fifteen years and whether your current job aligns with those visions.

Seek Feedback and Guidance

Seeking feedback from supervisors, mentors, and trusted colleagues can provide valuable insights into your strengths, areas for improvement, and potential career paths. This feedback can help you make informed decisions about your future. Feedback is a valuable source of information. It can provide you with an objective assessment of your performance and potential, helping you to make realistic and strategic career decisions.

Assess Job Satisfaction

Critically assessing your job satisfaction is crucial in determining whether a change is necessary. If you lack enthusiasm or feel unfulfilled in your current role, exploring new opportunities may be the key to revitalizing your career. Job satisfaction is an essential component of overall well-being, which is a major contributor to the quality of your performance. You are more likely to be engaged, motivated, productive, and excellent in your performance when you are satisfied with your job. If you are not satisfied with your current role or your current company, it may be time to leave.

Identify Transferable Skills

When considering an external move, you should identify transferable skills and experiences that make you a valuable candidate for a new role. Articulating how your current skills can be applied in a different context can strengthen your candidacy for external opportunities. By highlighting how your existing skills can add value to a new position, you can demonstrate your potential to succeed in a different capacity.

Evaluate The Proposed Company's Culture

Evaluate whether a new company's culture and values align with your own. A positive cultural fit can contribute to job satisfaction and a sense of belonging, making an external move with the right company the right career move at the right time. When you feel connected with your organization's culture, you are more likely to be engaged and committed to your role. Conduct in-depth research on the culture of any company you are considering before you accept an offer.

Conclusion

The fear of change can be a barrier to career advancement, causing you to miss out on valuable opportunities for growth and development. Staying too long in one job can lead to limited skill development, stagnation, and a narrow professional network. However, embracing change does not necessarily mean leaving the company altogether. An internal move to a new role within the same organization can be a strategic way to mitigate the career impact of staying too long. Weighing the decision to stay in your current role, making an internal move, or looking for a new role in a different company involves evaluating your career aspirations, seeking feedback and guidance, assessing your job satisfaction, evaluating your company's culture, and identifying your transferable skills. By taking a proactive approach to your career and embracing change when needed, you can ensure you stay on a path of continuous career advancement.

Leaving Too Soon

The advancements in AI may cause you to be concerned about your career prospects. This fear can lead you to consider leaving your current job prematurely in pursuit of what you believe to be greener pastures. However, making hasty career decisions without carefully considering the potential consequences can be detrimental to your long-term career growth.

The Fear of Missing Out

The fear of missing out on career advancement opportunities can be powerful and pervasive. As you watch colleagues or peers securing high-profile promotions, receiving impressive job offers, or achieving significant career milestones, you may feel a sense of restlessness and discontent in your current roles. You might also be concerned that attractive job opportunities will begin to dissipate. This fear-driven mentality can create an underlying sense of urgency, compelling you to seek better opportunities without thoroughly evaluating the potential impact of your decision.

This sense of restlessness can stem from a variety of sources. You may question your career trajectory and whether you're maximizing your skills and potential. You might compare your progress to your peers and wonder why you have yet to achieve similar recognition or advancement. These comparisons can be particularly intense in companies and/or industries where career milestones and achievements are highly visible and celebrated.

Moreover, there's the genuine concern that the window of opportunity for securing attractive job offers and career advancements may start to close. In rapidly evolving industries, the landscape can change swiftly. What seems like an abundance of opportunities today may appear scarce tomorrow. This fear-driven mentality can create an underlying sense of urgency, compelling you to seek better opportunities without thoroughly evaluating the potential impact of your decision.

While it's entirely natural to be motivated by the desire for career growth and success, it's essential to approach these feelings with a balanced perspective. The fear of missing out (FOMO) can be a potent force, but it should not be the sole driver of your career decisions.

Before you act on the impulse to look elsewhere, it's important to take a step back and consider a few key factors:

1. Self-Reflection: Take time to reflect on your career goals, values, and aspirations. What truly matters to you in your professional journey? What are your long-term objectives, and how do they align with your current role? Self-reflection can help you gain clarity on what you genuinely seek in your career.

2. Strategic Planning: Rather than making hasty decisions fueled by fear, develop a strategic plan for your career advancement. This plan should encompass short-term and long-term goals, skills development, and networking opportunities. A well-thought-out strategy will provide a roadmap for your career progression and prevent you from making an ill-conceived exit.

3. Informed Decision-Making: Gather as much information as possible when evaluating job offers or considering career changes. Conduct research, seek advice from mentors or trusted colleagues, and weigh the pros and cons. It's crucial to make informed decisions that align with your career objectives. Leaving too soon could land you in a company that is not right for you.

4. Embracing Patience: Recognize that career success is a journey and rarely follows a linear path. Some professionals achieve significant milestones early in their careers, while others progress more steadily. Embrace the idea that patience and persistence are valuable qualities in pursuing long-term success.

Premature Departures - A Double-Edged Sword

While leaving a job for what appears to be new horizons may seem like a promising path to career advancement, it can be a double-edged sword. On the one hand, a new job might offer exciting challenges and opportunities for growth. It may provide a chance to work with innovative technologies, collaborate with talented professionals, and potentially achieve higher compensation. On the other hand, hasty departures can lead to missed opportunities for building long-term relationships, acquiring valuable experience, and establishing a stable professional trajectory. By leaving too soon, you might fail to fully realize the potential of your current role, missing out on opportunities to create a meaningful impact within your current organization.

Limited Skill Development

Staying in a job for an optimal amount of time will allow you the opportunity to hone your skills and expertise in your chosen field at a place where you are doing well and have established a solid reputation. By investing time and effort into excelling in your current role, you can become a highly valued asset to your employers as you enhance your skills. Leaving too soon, however, can result in your walking away from the opportunity to develop expertise in your job function, potentially hindering future career growth.

The Appearance of an Unstable Trajectory

Frequent job-hopping can create a perception of instability on a professional résumé. Hiring managers and recruiters often view a series of short-term stays negatively, as it may indicate a lack of commitment and reliability. Instead of highlighting diverse skills and valuable experiences, a résumé peppered with brief job stints may raise concerns about the candidate's ability to commit to long-term projects and contribute effectively to an organization. Leaving prematurely may result in missed opportunities to make a lasting impact and build a strong track record of success. However, it is important to note that relatively short stints on your resume should not be detrimental to your career progression if you have a strong brand and your resume tells a good "story" indicating your career path and trajectory.

Reduced Network and References

Building strong professional relationships takes time, and leaving a job too soon may limit one's ability to develop a robust network within the company. Over time, professionals forge connections with team members, mentors, and company leaders, providing them with a support system and access to valuable opportunities. Additionally, short tenures can make securing strong references from former colleagues and supervisors challenging.

Assess Your Current Role

Before making any decisions, you should thoroughly evaluate your role. Consider the challenges you face, the growth opportunities available, and the potential for advancement within your current organization. Conducting a thorough role assessment can help you determine whether your current job aligns with your career goals and whether there are opportunities to learn, contribute, and grow within your current organization.

Consider the New Opportunity

When presented with a new job offer or advancement opportunity, it is essential to conduct thorough research. Consider factors such as the prospective company's culture, growth potential, job responsibilities, and the alignment of the role with your career goals. By seeking insights from current or former employees, exploring the company's reputation in the industry, and understanding the challenges and opportunities associated with the new role, you can make more informed decisions about whether the potential benefits outweigh the risks.

Seek Advice

You will benefit from seeking guidance from a career coach. Your coach will provide valuable insights and an objective perspective on potential career moves. Your coach can walk you through the best way to assess the pros and cons of new opportunities and help you strike a balance between following your ambitions versus the potentially negative impact of leaving your job too soon.

Weighing the Personal Impact

You should also take into account the personal impact of a job change, such as the potential for relocation, work-life balance, and family considerations. Achieving career growth without sacrificing personal well-being is crucial for long-term career satisfaction. All career moves can be stressful, even under the best circumstances. The cost to your personal life must be included in your evaluation.

Identify Red Flags

Watch out for red flags in both your current job and your potential new opportunity. These red flags could include excessive turnover within an organization, unclear expectations for your role, or a lack of alignment with your personal values. By being vigilant about potential warning signs, you can make informed decisions about the risk of moving to a job or the risk, if any, of staying where you are.

Seek Internal Advancement

If the desire for career advancement arises from a perceived lack of growth opportunities in the current role, consider exploring potential paths for internal advancement within your current organization. By initiating conversations with supervisors or HR personnel about potential growth opportunities, you can demonstrate your commitment to your organization and your willingness to take on new challenges within the company.

Conclusion

Leaving a job too soon out of fear of missing out on career advancement opportunities can have significant consequences on your long-term career growth. While remaining proactive in pursuing career advancement is essential, you must also weigh the pros and cons of job changes and consider the potential career impact that may result from hasty decisions. You can make informed career decisions that support your long-term growth and success by evaluating your career goals, assessing your current roles, and carefully considering new opportunities. The fear of missing out on career opportunities is a powerful motivator. It should never drive impulsive decisions. Instead, use it as a catalyst for self-reflection, strategic planning, and informed decision-making. Your career journey is unique, and by approaching it with a thoughtful and balanced perspective, you will make the best decisions for you.

Doing Too Much

The advancements of AI and automation will undoubtedly transform the workforce, enhancing productivity and streamlining processes. However, this technological revolution has also given rise to apprehensions among professionals regarding the potential of being replaced by AI. The fear of AI taking over the jobs that humans have traditionally performed can be overwhelming.

Indeed, the fear of being replaced by AI is just one example of how work-related fears can drive employees to do too much at work. There are various other fears and anxieties that can result in a similar pattern of overwork and its associated consequences:

1. Fear of Job Insecurity: In addition to AI, concerns about job stability due to factors like economic downturns, company reorganizations, or industry disruptions can induce excessive work efforts. Employees might take on extra responsibilities to demonstrate their value to the organization, fearing their jobs could be at risk.

Valerie Capers Workman, Esq.

2. Fear of Peer Comparison: The fear of falling behind or not measuring up to colleagues can be a potent motivator for overworking. If employees perceive their peers as achieving more or progressing faster in their careers, they may feel the need to work incessantly to match or exceed those standards.

3. Fear of Missing Out (FOMO): This fear extends beyond career advancement to opportunities within the workplace. Employees may fear missing out on exciting projects, training, or experiences and may overcommit to ensure they are part of everything, often at the cost of their well-being.

4. Fear of Negative Evaluation: A strong desire to avoid criticism or negative evaluations from supervisors or peers can lead to overworking. Employees may go to great lengths to ensure their work is perceived as flawless, fearing the consequences of any perceived shortcomings.

5. Fear of Economic Insecurity: Concerns about financial stability and the need to support one's family can drive individuals to overwork. This fear can be especially pronounced when the cost of living is high and job opportunities are competitive.

6. Fear of Career Plateau: Employees may fear that their careers have plateaued, with limited opportunities for growth or advancement. This fear can lead to a constant drive to take on more responsibilities or seek new challenges, even if it means overextending oneself.

These various fears can create a constant state of heightened stress and anxiety. While they can initially drive increased efforts and strong performance, over time, they often result in burnout, decreased job satisfaction, and potential physical and mental health issues.

To address these fears and their associated behaviors, it is essential to cultivate self-awareness and seek a healthy work-life balance. Recognizing and addressing these fears can lead to a more sustainable and fulfilling work experience.

The Potential Psychological Impact of AI

The fear of being replaced by AI could lead to heightened stress, anxiety, and a constant drive to overperform. You might feel compelled to put in extra hours, taking on more tasks than you can handle in an attempt to prove your value and indispensability. This fear-driven work mentality could result in you sacrificing your personal time and well-being, leading to an increased risk of burnout and diminished job satisfaction. Moreover, the constant exposure to news headlines about AI advancements and the potential impact on the workforce can perpetuate a sense of uncertainty and insecurity.

The Dangers of Overwork and Burnout

Overworking and neglecting self-care can have severe consequences on physical and mental well-being. Burnout, a state of chronic stress, can lead to emotional exhaustion, reduced job performance, and a sense of detachment from work and colleagues. Burnout not only affects you but also has broader implications for the organization, leading to increased absenteeism, reduced productivity, and higher employee turnover.

Understanding the Limits of AI

It is essential for you to recognize that while AI can automate repetitive tasks and handle large amounts of data, it has its limitations. AI lacks human qualities such as creativity, emotional intelligence, and critical thinking, making human skills invaluable in many aspects of the workplace. For instance, while AI-driven

183

algorithms can analyze data and identify patterns, human intuition and creativity are necessary to make sense of complex and ambiguous situations.

Understanding these limitations can help you see AI as a tool to augment your abilities rather than a direct competitor. By recognizing the unique strengths that humans (you!) bring to the table, you can begin to shift your perspective from viewing AI as a threat to embracing it as a collaborator.

AI as a Collaborator

Instead of viewing AI as a threat, you should see it as a collaborative tool that complements your abilities. AI can handle routine and data-intensive tasks, freeing you to focus on tasks that require human ingenuity and emotional intelligence. For example, in customer service, AI-powered chatbots can efficiently handle repetitive inquiries, allowing human agents to engage in more complex and empathetic customer interactions. This collaborative approach ensures that professionals can leverage AI to enhance their productivity and efficiency rather than feeling displaced by it. Alleviating work-related fears without subjecting oneself to unnecessary overwork or compromising mental health is essential for maintaining a healthy work-life balance and overall well-being.

EFFECTIVE STRATEGIES TO MANAGE FEARS:

1. Self-Assessment and Awareness

a. Start by understanding your fears and anxieties. Reflect on what triggers them and how they impact your behavior.

b. Recognize that some level of fear and stress can be motivating, but excessive anxiety can lead to burnout.

2. Set Realistic Goals and Boundaries

a. Establish clear, achievable career and personal goals. Setting realistic expectations can help reduce anxiety about falling short.

3. Continuous Learning and Skill Development

a. Instead of overworking, focus on continuous learning and skill development. Invest time in improving your skills and knowledge to stay competitive.

b. Seek opportunities for professional development within your current role or organization.

4. Seek Support and Feedback

a. Engage in open and honest communication with your supervisor, colleagues, or mentors.

b. Welcome constructive feedback as a tool for growth rather than fearing criticism.

5. Time Management and Prioritization

a. Learn effective time management techniques to prioritize tasks and optimize productivity without excessive work hours.

b. Use tools like project management apps to help organize and manage your workload.

6. Stress Management and Well-Being

a. Incorporate stress management techniques into your daily routine, such as prayer, meditation, yoga, mindfulness, or exercise.

b. Prioritize self-care and well-being, including adequate sleep, a balanced diet, and regular exercise.

7. Seek Professional Help, if needed

a. If work-related fears significantly impact your mental health or well-being, consider consulting a therapist or counselor who specializes in workplace stress and anxiety.

b. Employee assistance programs (EAPs) at your company may offer confidential counseling and support services.

8. Evaluate Job Fit and Career Goals

a. Periodically assess whether your current job aligns with your long-term career goals and values.

b. Consider seeking new opportunities that better align with your aspirations, if necessary.

Conclusion

The integration of AI in the workplace has the potential to spark fears leading to overwork and burnout. However, it is crucial to recognize that AI is not a replacement for human skills but rather a complementary tool. Strategies such as setting boundaries, delegating tasks effectively, and prioritizing self-care are essential in avoiding burnout while embracing AI.

Imposter Syndrome[32]

Imposter syndrome is a common experience many professionals encounter at some point in their careers. It can sometimes lead to self-doubt and hinder your confidence. However, it's essential to remember that these feelings are not uncommon, and they certainly don't have to define your career trajectory. While it's natural to question your abilities from time to time, it's equally important to recognize your strengths and achievements. Instead of allowing imposter syndrome to hold you back, consider it an opportunity for personal growth and development.

Understanding your worth, embracing your skills, and nurturing your emotional intelligence (EQ) can help you navigate through these feelings. By doing so, you can seize career opportunities as they arise.

---

[32] The term "Imposter Syndrome" was coined by psychologists Dr. Pauline R. Clance and Dr. Suzanne A. Imes in their research paper published in 1978. They conducted a study focused on high-achieving women and their experiences of self-doubt and feeling like "impostors" despite their evident success.

Valerie Capers Workman, Esq.

It's worth noting that imposter syndrome affects people differently. Some individuals may become hesitant in career-defining moments, while others might adopt an overcompensating attitude. Both responses can be managed through self-awareness and seeking support and guidance when needed.

Imposter syndrome is a psychological challenge that many successful professionals have faced and overcome. With the right mindset and strategies, you can manage and even use these feelings to fuel your personal and professional growth. It's about acknowledging the challenge without allowing it to define your path.

Understanding Imposter Syndrome

Imposter syndrome is a real mental phenomenon based on irrational feelings of self-doubt, insecurity, and inadequacy despite tangible evidence of your competence. Regardless of their achievements, many professionals struggle with a persistent fear that "someone" will discover they do not belong in certain spaces. Imposter syndrome causes a variety of unproductive behaviors, including:

1. **Perfectionism:** Setting unattainably high standards for themselves, leading to excessive self-criticism and reluctance to take on new challenges for fear of failure.
2. **Attribution of Success:** Believing that circumstances beyond their control rather than their skills and efforts are the reason for their success, also causing them to downplay their achievements and fail to take credit where credit is due.
3. **Fear of Evaluation:** Fear of being evaluated or judged by others, leading to anxiety about their performance and a tendency to avoid situations that might highlight their perceived inadequacies.

4. **Overworking:** To prove their worth, professionals may overcompensate by working excessively, leading to burnout and perpetuating the cycle of feeling inadequate.

The Impact of Imposter Syndrome on Your Career

1. **Missed Opportunities:** Professionals battling imposter syndrome might hesitate to seize opportunities for career advancement, fearing they are not qualified or deserving. This can lead to missed chances to contribute meaningfully to career-making projects and initiatives.
2. **Stagnation:** Professionals succumbing to imposter syndrome may refrain from voicing their opinions and ideas in strategic discussions, leading to a lack of influence and relegating them to passive roles as observers rather than active contributors.
3. **Limiting Growth:** Imposter syndrome can hinder professional growth, causing individuals to opt for safer known paths rather than embracing risks and exploring new avenues for learning and development.
4. **Underestimating Potential:** Professionals who underestimate their potential may inadvertently limit their contributions to at work, perpetuating a cycle of unrealized opportunities.

Conquering Imposter Syndrome

Conquering imposter syndrome is not only a personal imperative but also a strategic necessity in the age of AI. Professionals can undertake several steps to overcome these feelings of inadequacy and position themselves as formidable contributors at the proverbial table. Here are the steps I followed when I fought and conquered my bout with imposter syndrome earlier in my career.

Valerie Capers Workman, Esq.

1. **Acknowledge Feelings:** The first step is to recognize and accept the presence of imposter syndrome. Acknowledging these feelings will diminish their power and pave the way for professional self-reflection.

2. **Celebrate Achievements:** Actively recognize and celebrate your achievements, attributing and connecting your successes directly to their skills, efforts, and talents rather than dismissing them as coincidences.

3. **Seek Mentorship:** Engaging with mentors who can offer guidance and share their experiences with Imposter Syndrome can provide invaluable perspective and encouragement.

4. **Focus on Learning:** Embracing a growth mindset is essential. View challenges and setbacks as opportunities for learning and growth, dispelling the notion that you must be perfect from the start.

5. **Practice Self-Compassion:** Cultivating self-compassion involves treating yourself with the same kindness and understanding you would offer to a friend facing similar challenges.

6. **Set Realistic Goals:** Set achievable goals that allow for incremental progress, reducing the pressure to attain unrealistic levels of perfection.

7. **Embrace Failure:** Understanding that failures are inherent in growth and innovation can help you navigate setbacks with resilience and a willingness to adapt.

8. **Visualize Success:** Practicing positive visualization techniques will help you imagine yourself succeeding in challenging situations, boosting your confidence and self-esteem.

9. **Focus on EQ:** Building emotional intelligence is paramount. You should develop self-awareness, self-regulation, social skills, empathy, and motivation, enabling you to navigate

complex interpersonal dynamics confidently. Therapy is a great resource.

Conclusion

In the age of AI, where professionals are faced with unprecedented opportunities to contribute to strategic decisions, conquering imposter syndrome is not just a personal journey but a strategic imperative. The false constructs of inadequacy must be silenced to unlock the full potential of your skills, abilities, and EQ. As AI reshapes industries, you must proactively embrace opportunities rather than retreat from them due to imposter syndrome. By recognizing your worth, celebrating your achievements, seeking mentorship, and cultivating a growth-oriented mindset, you will allow yourself to confidently stride into your continued career progress.

Unconscious Bias[33]

As workplaces become increasingly more diverse, your ability to acknowledge and manage the impact of your unconscious bias on your work interactions will help to ensure your ability to advance in your career.

It is important to note that every human being has an unconscious bias, the significance of which is directly proportional to your level in your organization.

Understanding Unconscious Bias

Unconscious bias refers to automatic and unintentional mental shortcuts people make when processing information and making decisions. These biases stem from ingrained societal and cultural norms, personal experiences, and exposure to media, and they can significantly influence perceptions, actions, and interactions without

---

[33] Greenwald, A & Banaji, M (1995). Implicit social cognition: Attitudes, self-esteem and stereotypes. Psychological Review 1995, Vol 102, No. 1, 4-27. American Psychological Association

conscious awareness. Unconscious biases can manifest in various forms:

- **Confirmation Bias**: People tend to seek out and interpret information confirming their existing beliefs, potentially disregarding alternative perspectives.

- **Stereotyping**: Individuals assign generalized characteristics to groups based on preconceived notions, leading to inaccurate assumptions.

- **Halo Effect:** Positive or negative traits attributed to one aspect of a person influence judgments about their overall character.

- **Affinity Bias**: People favor others with similar backgrounds, experiences, or traits, often leading to unequal opportunities.

- **Attribution Bias:** Successes of in-group members are attributed to skills, while failures are seen as exceptions, whereas the reverse is true for out-group members.

The Impact of Unconscious Bias in the Age of AI

In the evolving landscape of AI-driven industries, unconscious bias can be particularly detrimental due to its potential to hinder collaboration, impede innovation, and foster inequitable practices. The implications are extensive:

1. **Recruitment and Hiring**: Unconscious bias can lead to biased candidate evaluation and skewed hiring decisions, resulting in missed opportunities to hire the best candidates for open roles **and** squandering the chance to diversify teams with varied skills and perspectives.

2. **Leadership and Management:** Professionals unaware of their biases might unwittingly marginalize team members, stifle contributions, and undermine cohesion.

3. **Innovation and Creativity:** Unconscious bias can suppress diverse viewpoints, stifling creative problem-solving and diminishing the potential for groundbreaking innovations.

4. **Workplace Dynamics:** Negative interactions stemming from unconscious bias can erode trust, hinder collaboration, and foster a divisive environment.

5. **Career Advancement**: Professionals who overlook the impact of unconscious bias may inadvertently hold back underrepresented employees, stifling their growth potential.

Conquering Unconscious Bias

To thrive in the age of AI and leverage its transformative power, you must recognize, address, and manage your unconscious biases.

1. **Educate and Raise Awareness**: You must actively seek education on unconscious bias, its forms, and its impacts. Training sessions, workshops, and reading materials can facilitate a deeper understanding of these biases.

2. **Practice Self-Reflection:** Regularly examining your thoughts, actions, and decisions can help you identify instances of unconscious bias. Feedback from colleagues can provide valuable insights.

3. **Cultivate Empathy**: Developing empathy—the ability to cultivate empathy—will enable you to relate to others'

experiences, thereby reducing biases and promoting inclusive interactions.

4. **Diversify Networks:** Expanding your network to include individuals from different backgrounds fosters exposure to diverse perspectives, broadening your viewpoints.

5. **Implement Structured Decision-Making:** Implementing structured data-driven decision-making helps mitigate the influence of unconscious biases on critical choices.

6. **Promote Inclusive Leadership:** Leaders must set an example by acknowledging their biases, actively seeking feedback, and championing an inclusive culture that embraces Equal Access.

7. **Embrace Feedback:** Constructive feedback from colleagues can provide valuable insights into instances of unconscious bias that professionals may not have recognized otherwise.

8. **Implement Unbiased Language**: Using inclusive language in communications and interactions can counteract subconscious biases that may perpetuate inequities.

9. **Champion Equity:** You can advocate for equitable representation in decision-making roles.

Conclusion

The ability to manage biases is an essential skill for career advancement. Failure to address unconscious bias perpetuates inequities and impedes career advancement by limiting your ability to effectively navigate complex interpersonal dynamics – a skill that is particularly critical in the age of AI.

"There Is No Fate but What We Make for Ourselves"

~ John Connor

The rise of AI in the workplace has been marked by excitement, innovation, and, for many, a degree of apprehension. The fear that machines will replace human jobs has become a prevalent concern. But as I have explored throughout this book, this fear, although not entirely baseless, can be mitigated and even turned into an opportunity.

AI is not necessarily an adversary to human workers; it can be an ally. Automating mundane tasks frees us to focus on more complex and creative aspects of our work. By understanding and leveraging AI's capabilities, we can augment our skills and even enhance our workplace value. The dynamic nature of technology means that what's relevant today might become obsolete tomorrow. The key to thriving in an AI-driven environment is continuous learning. This means:

1. **Understanding emerging technologies** – stay abreast of the latest advancements in AI and related fields.
2. **Upskilling and reskilling** – regularly assess your skills against industry needs and take proactive steps to bridge any gaps.

3. **Adopting a growth mindset** – view changes not as threats but as opportunities to grow and learn.

In a world where AI is becoming more integral, human attributes like creativity, empathy, and strategic thinking are increasingly valuable. Machines cannot replicate these attributes. By enhancing these competencies, you will have more job opportunities than you could imagine.

The age of AI is not a doomsday scenario for human professionals. It is a call to action. The future of work is changing, but it is not something that is happening to you—it is something you can actively shape. By keeping your skills current, seeking continuous learning, focusing on human-centric attributes, and building resilience, you can survive and thrive in the AI-driven landscape. Using my career path as a guide, you can make multi-dimensional moves that will enable you to continue advancing in your career with strategic agility. The future is not something to fear but a canvas on which you have the power and the tools to paint your destiny. The integration of AI in the workplace is not a closed door but a gateway to uncharted territories filled with potential. Your fate in this new era is not predetermined; it's what you choose to make for yourself. And with the right mindset, knowledge, and game plan, your future will be one of great promise and fulfillment.

Now that you are armed with the art and science of career advancement, you have the power of Quantum Progression to propel your career as far as you want it to go.

"Open The Pod Bay Doors. Please, HAL!"

Valerie

About The Author

Valerie Capers Workman is the former Vice President, People for Tesla and former direct report to Elon Musk, a business executive, and the current Chief Legal Officer at Handshake. Recognized as a hyper-scale expert, Valerie has guided iconic consumer brand growth and new market entry for Tesla, FOCUS Brands (owner of beloved F&B brands including Cinnabon, Auntie Anne's Pretzels and Moe's Southwest Grill), Realogy (owner of real estate powerhouses including Century 21, Coldwell Banker and Sotheby's), and Wyndham Hotels & Resorts (owner of iconic brands including Ramada, Days Inn, Howard Johnson's and Microtel), fusing 20 years of corporate governance, operational, M&A, global expansion, and human capital management strategies.

During Valerie's four years with Tesla, she was Head of Compliance, paving the way for the construction of the Gigafactory in Shanghai. In December 2019, Elon Musk designated Valerie to lead the U.S., EMEA, and non-China APAC People functions, including

Valerie Capers Workman, Esq.

HR and Recruiting. Three months after taking on this role, COVID-19 hit the U.S., and Valerie helped lead the company's growth from 47k to over 100k employees while co-introducing innovative employee protection programs with Tesla's VP, Environmental Health, Safety & Security. In July 2020, Valerie was promoted to VP, People for Tesla. During her tenure with Tesla, Valerie became known as the company's "Top Diversity Advocate" and led the creation of numerous programs to enhance the work lives of employees. Under her leadership, Tesla published its first-ever DEI Report and replaced its "Anti-Handbook Handbook" with a comprehensive online Employee Guidebook. Valerie's work included enhancing pay equity processes, expanding mental health benefits, expanding access to reproductive options, supporting employees' access to the family planning decisions that were right for them, and leading the creation and rollout of Tesla's company-wide "Respectful Recharge" training.

Currently, she serves as the Chief Legal Officer for Handshake, where she is the Leader of the Legal, Global Support, and Global Trust & Safety teams. Handshake's community includes over 13 million students and alumni around the world from 1,400 educational institutions, including four-year colleges, community colleges, boot camps, and 300+ minority-serving institutions. The platform connects up-and-coming talent with 850,000+ employers - from Fortune 500 companies to thousands of public school districts, healthcare systems, nonprofits, and more.

Valerie is admitted to practice law in New York, Texas, and Washington, DC. She resides in the greater Austin, Texas area with her wonderful husband and their three amazing tech-forward sons. She is a graduate of the S.I. Newhouse School of Public Communications at Syracuse University and St. John's University School of Law. Valerie is a proud member of Alpha Kappa Alpha Sorority, Incorporated and Black Women on Boards.